Veteran's Stress, Trauma and Synapse Rehabilitation

Maxwell Bennett

Veteran's Stress, Trauma and Synapse Rehabilitation

Springer

Maxwell Bennett
Department of Physiology
University of Sydney
Sydney, NSW, Australia

ISBN 978-981-95-9593-8 ISBN 978-981-95-9594-5 (eBook)
https://doi.org/10.1007/978-981-95-9594-5

This Springer imprint is published by the registered company Springer Nature Singapore Pte Ltd.
The registered company address is: 152 Beach Road, #21-01/04 Gateway East, Singapore 189721, Singapore

For

Admiral Chris Barrie

Preface

Overview

This book is a multidisciplinary work attempting to inform graduate students, clinical and preclinical researchers as well as mental health professionals as to the best way forward in the study and treatment of combat-related post-traumatic stress disorder (PTSD). The kind of PTSD that arises from combat and its consequences for the life of the individual and their family is first considered (Part II). Then the extent to which psychotherapies and pharmacotherapies, as well as social support, can ameliorate PTSD sets the stage for the main hypothesis under consideration (Part III). The main hypothesis guiding this work is next spelt out: the suffering arises from a loss of grey matter in core parts of the brain related to PTSD behaviour, and that this is due to the degeneration of synapses in specific neural networks of the brain consequent on combat exposure (Part IV). The principal issue is then to rehabilitate these synapses in the neural networks consequent on combat exposure, for which methods that promote the synthesis and release of the brain's endogenous synapse growth factor are of major interest. The most efficacious of these methods are shown to be transcranial magnetic stimulation, neurofeedback and psychedelics (Parts V and VI). Finally, the question arises as to how the actual distortion of consciousness, including dissociative consciousness, arises in PTSD for which a second hypothesis is offered in detail related to specific cortical synapses (Part VII): the distortions arise from failure of synaptic integration of cortical and sensory experiences, leading to dissociative symptoms such as derealization and depersonalization, requiring for their remedy the rehabilitation of particular synapses on the processes of the largest neurons in the brain.

Details

The first parts of this work are structured around the narrative concerning the experiences of an Australian First World War veteran, called Pop, who had fought in the terrible Battle of the Somme in France in 1916[1] (Chap. 1). The origins in battle that gave rise to Pop's 'shell shock', later called PTSD, are given in detail from his diary, providing the basis in succeeding chapters for how different experiences in combat relate to his condition. It is first necessary to rigorously define PTSD in its clinical manifestations and to distinguish it from major depressive disorder (MDD) with which it is frequently comorbid. The American Psychiatric Association has largely guided the development of criteria that need to be met for a diagnosis of PTSD (Chap. 2). The extent and intensity of combat experiences, like those of Pop, are then related to his lifetime and current experiences of PTSD. Observing death, injury and experiencing personal threats are highlighted as major stresses, leading to vivid re-experiencing intrusions among the clusters of PTSD symptoms (Chap. 3). Another PTSD symptom cluster arising from battle is that of hyperarousal, closely associated with suicidality and high-risk behaviour, all of which are exacerbated by poorly managed transitions to civilian life (Chap. 4).

Tragically, it is not only the veteran with PTSD that suffers from the consequences of battle but also his family and friends. The hyperarousal and numbing symptom clusters of PTSD are associated with aggressive behaviour, intimate partner violence, substance abuse and related crimes (Chap. 5). Particularly significant is the impact of the veteran's behaviour on his children, with a high probability of these being subjected to emotional neglect and abuse, as recorded in relation to Pop's young daughter. The result is the intergenerational transfer of the consequences of trauma and its mental health sequelae (Chap. 6).

After this depressing litany of the consequences of battlefield-induced PTSD there follows consideration of the best psychotherapeutic and pharmacotherapeutic methods available for ameliorating these consequences. The core psychotherapies considered are exposure, cognitive restructuring, bilateral stimulation and narrative integration therapies. The relative effectiveness of each of these is considered (Chap. 7), with emphasis that therapies for a PTSD patient must be tailored to the particular PTSD symptom cluster that is paramount in the patient's suffering. Turning to the pharmacotherapies, it is the relatively new non-classical atypical psychedelics, such as ketamine and MDMA (3,4-methylmediocymethamphetamime), that are highlighted. Indeed, it is emphasized that with careful clinical oversight these psychedelics have powerful effects in relieving PTSD symptom clusters (Chap. 8). Overlying all of these treatments is the necessity for social support for the returning veteran, especially emotional support coupled with very positive social interactions. It is support from peers, that is from other veterans, that is especially efficacious

[1] In this battle the number of Australian soldiers killed as a percentage of the then Australian male population was about the same as that of the French who were defending their own country.

(Chap. 9), for it enhances in a quantitative way the response to psychotherapeutics and pharmacotherapeutic treatments.

The next chapters begin the analysis of brain changes underlying PTSD symptoms, an investigation necessary for the development of new treatments. It introduces the hypothesis that the key to understanding PTSD lies in determining the changes that occur during battle at the level of the synaptic connections between neurones that constitute the brain's neural networks.[2] Principal among these are the default mode network underlying rumination, the central executive network supporting high-level cognitive functions and the salience network for detecting and prioritizing stimuli that guide behaviour. The processes of neurons and their synaptic connections dominate the grey matter volume, so the loss of this volume in these three networks points to their decline in efficiency, which in turn is associated with particular combat experiences and specific PTSD symptoms (Chap. 10).

How is this loss of grey matter and hence of neural network synaptic functions to be restored following combat experience? Pharmacotherapies are at least partly effective in restoring grey matter (a surrogate, as we have seen, of synaptic connectivity) of neural networks, with some acting to increase connectivity in the hippocampus and prefrontal brain regions. Among the neurotherapies, neurofeedback enhances both frontal cortex and temporal cortex grey matter in patients. There is a range of therapeutic interventions for PTSD patients that lead to at least partial return of grey matter and hence of synaptic connections (Chap. 11).

The key to the rehabilitation of synapses following combat and hence of grey matter is the use of therapeutic agents that enhance endogenous brain-derived growth factors for restoring the lost synapses, the principal one of which is Brain-Derived Neurotrophic Factor (BDNF) (Chap. 12). Therapeutic interventions enhance BDNF signalling, synaptogenesis and hence grey matter, leading to clinical improvements. The agents that have been identified to this date that enhance BDNF are the psychedelics such as ketamine as well as Transcranial Magnetic Stimulation (TMS) (Chap. 13).

The loss of grey matter due to battle, reflecting the loss of synaptic connectivity, needs to be restored to ameliorate PTSD, especially in the default mode, cognitive and salient neural networks of the brain (Chap. 14). The extent to which the normal functioning of these networks in PTSD patients can be effected using trauma-focussed cognitive behaviour therapy, ketamine, TMS and EEG-directed neurofeedback is considered in detail. Of these, TMS and neurofeedback have the highly significant effect of reducing PTSD symptoms of between 50% and 70% (Chap. 15).

Reviewing the evidence for all of the different therapies leads to the conclusion that the ones that Pop needed after the Battle for the Somme were TMS, EEG-guided neurofeedback and ketamine (Chap. 16). These have received emphasis in this work, and are being used by a growing number of clinicians, but are still 'off-label', that is, for inexplicable reasons, are not approved by the Food and Drug

[2] There are about 100 billion neurones in the brain, which connect to each other via these one micrometre diameter synapses, of which each neurone has on average about 10,000, giving altogether 1000 trillion synapses, indicative of the immense complexity of the brain's neural networks.

Administration in the United States or the Therapeutic Goods Administration in Australia. This is reminiscent of the story of John Cade, the Australian who returned from battles in the Pacific during the Second World War, to discover lithium for the treatment of bipolar disorder in 1949, the first and still highly successful treatment for any mental illness.[3] This was accepted by Australia for that purpose but not by the Food and Drug Administration for 20 years after its discovery.

What actually occurs in the brain during the distortions of consciousness that PTSD patients experience (Chap. 17)? A grievous consequence for those who go into battle on behalf of their countries, like Pop did (Chap. 1), is that they may lose synapses in particular parts of their brains that lead to delusional states. It is hypothesized here that it is the loss of synapses on particular processes of the largest neurons in the cortex (their apical tuft dendrites) that leads to delusional states of consciousness. The rehabilitation of these synapses should be a focus of PTSD research in the support of veterans to lead normal lives again after return from battle.

Reference

De Moore G, Westmore A (2016) Finding sanity: John Cade, lithium and the taming of bipolar disorder. Allen & Unwin, Australia

Sydney, NSW, Australia Maxwell Bennett

[3] de Moore and Westmore (2016).

Also by Maxwell Bennett

Neuroscience of Synapses

'Autonomic Neuromuscular Transmission' (1972)
'The Idea of Consciousness: Synapses and the Mind' (1997)
'History of the Synapse' (2001)
'The Neurobiology and Biophysics of Synapses: A Memoir' (2020)

Neuropsychiatry

'Virginia Woolf and Neuropsychiatry' (2013)
'Stress, Trauma and Synaptic Plasticity' (2018)
'Childhood Stress, Trauma and Synapse Loss' (2024)

Neurophilosophy

'Philosophical Foundations of Neuroscience' (2003) Second Edition (2023)
with P. Hacker

'Neuroscience and Philosophy: Brain, Mind and Language' (2006)
with D. Dennett, P. Hacker, J. Searle

'History of Cognitive Neuroscience' (2008)
with P. Hacker

'The Representational Fallacy in Neuroscience and Psychology' (2024)
with P. Hacker

On Universities

'The Search for Knowledge and Understanding' (2019)
'The Search for Truth: History and Future of Universities' (2022)

On Founding Mental Health Institutes

'Founding the Brain & Mind Research Institute (Sydney)' (2002)'Founding the Mind and Neuroscience – Thompson Institute (Maroochydore)' (2012)

Acknowledgements

My wife of 62 years, Gillian, has guided and supported me through all my efforts to understand brain function in relation to psychiatric diseases over four volumes constituting a treatise on the neuroscience of stress and trauma.[1] My wonderful editor, Linda Nix, has not only carried out the usual editorial tasks but has also been a constant source of advice in bringing clarity to a complex interdisciplinary work. Brilliant colleagues at mental health institutes I have been involved with to ameliorate mental suffering have been instrumental in keeping me active in brain research at 88. These are *The Brain & Mind Research Institute* (now Centre at Sydney), *The Mind & Neuroscience Thompson Institute* (now Thompson Institute at Maroochydore) and the clinical not-for-profit *Brain & Mind Healthcare Hubs*. Both the writing of books such as this one and the establishment of institutes required sustained support from senior members of the community. In this regard outstanding generosity has been received in recent times from a great advocate for the well-being of veterans, Admiral Chris Barrie, to whom this book is dedicated.

[1] Of the four volumes, *Stress, Trauma and Synaptic Plasticity* (with J Lagopoulos) provides molecular insights into brain changes accompanying stress and trauma. *Virginia Woolf and Neuropsychiatry* begins with the narrative of her life involving early sex abuse by her half-brothers before examining how the accompanying brain changes led to suicidality and her eventual suicide. Next, *Childhood Stress, Trauma and Synapse Loss* begins with the narrative of an abused child and traces how, in particular, emotional abuse of the child is a principal driver of deleterious personality changes and major depression through the loss of synaptic connections in their brains. Finally, the present work, *Veteran's Stress, Trauma and Synapse Rehabilitation*, shows, following a narrative of a veteran's battlefield experiences, how these can lead to loss of synapses in particular neural networks of their brains that underlie PTSD, even though the veteran is not directly wounded; this work ends with identifying what treatments are available that will rehabilitate these synapses and return the veteran to a normal life.

Contents

Part I Vignettes

1 Vignettes of Australian Veterans and 'Shell Shock' in World War I . 3
1.1 Introduction 3
1.2 Vignette 1: Origins and Enlistment 4
1.3 Vignette 2: The Battle from Sausage Valley and the Chalk Pit Road to the Capture of Pozières Village. 5
1.4 Vignette 3: The Battle for the Old German Lines, North–North-East of Pozières 8
1.5 Shell Shock and Its Medical Treatment in the World War I. 10
1.6 Vignette 4: Post-war Nightmares and Violent Behaviour 12
1.7 Vignette 5: Inability to Find Work 14
1.8 Vignette 6: Inability to Remain Employed 15
1.9 Vignette 7: Effect of PTSD on Family. 16
References. 18

Part II Veterans' PTSD Following Different Combat Experiences

2 Clinical Identification of PTSD 21
2.1 DSM-5 Criteria for PTSD Symptom Clusters 21
2.2 Application of the DSM-5 Criteria to the Experiences of an Australian Veteran. 23
2.2.1 Intrusion (PTSD Symptom Cluster B). 23
2.2.2 Avoidance (PTSD Symptom Cluster C) 24
2.2.3 Negative Alterations in Cognition and Mood (PTSD Symptom Cluster D) 24
2.2.4 Alterations in Arousal and Reactivity (PTSD Symptom Cluster E) 25
2.3 Measures Used to Assess PTSD Symptoms 26
2.3.1 The Posttraumatic Stress Disorder Checklist for DSM-5 (PCL-5) 26
2.3.2 The Clinician-Administered PTSD Scale (CAPS) 27

2.3.3 Comparing PCL-5 and CAPS ... 30
2.4 Comorbidity of PTSD and Major Depressive Disorder (MDD) ... 30
2.4.1 MDD Symptom Overlap with DSM-5 PTSD Symptoms .. 30
2.4.2 MDD Symptom Overlap with Dysphoria Models of PTSD ... 32
2.4.3 Impact of Trauma Type on Comorbidity ... 32
References ... 36

3 Combat Experience Leading to PTSD ... 39
3.1 Combat Exposure Scale ... 39
3.1.1 Dose–Response Relationship of Combat Exposure with PTSD ... 40
3.1.2 Relationship of Combat Exposure with Duration of PTSD... 41
3.2 The Relationship Between Critical War-Zone Experiences and Mental Health Problems, Including PTSD ... 42
3.3 Combat Experience Scale ... 42
3.3.1 Correlations Between Combat Experiences and Psychopathologies ... 42
3.3.2 Correlations Between Combat Experiences and PTSD Symptom Clusters ... 45
3.4 Application of the Combat Experience Scale to the Experiences of an Australian Veteran ... 46
3.4.1 Violent Combat ... 46
3.4.2 Proximity to Wounding or Death ... 47
3.4.3 Combat Injury ... 48
References ... 49

4 Suicidality Comorbid with PTSD ... 53
4.1 PTSD and Suicidality: An Overview ... 53
4.2 The Impact of Deployment-Related Trauma ... 55
4.3 Impact of Psychosocial Factors ... 57
4.3.1 Impulsivity ... 57
4.3.2 Transition to Civilian Life ... 57
4.3.3 Mental Health Issues ... 57
4.3.4 Lack of Social Support ... 57
4.3.5 Substance Abuse ... 58
4.4 Conclusion ... 58
References ... 58

5 Criminality Comorbid with PTSD ... 63
5.1 Veteran PTSD and Criminality ... 63
5.2 Alcoholism as an Amplifying Risk Factor ... 64
5.3 Exposure to Combat as an Amplifying Risk Factor ... 64
5.4 Synthesis of Findings ... 64
References ... 65

6 Family Impact of a Veteran's PTSD 67
6.1 Children 67
6.1.1 PTSD Symptoms of a Parent and Abuse of Child 67
6.1.2 Consequential Effects on Child 68
6.2 Partners 69
6.3 Implications 71
References 71

Part III Ameliorating Veterans' PTSD with Psychotherapies and Pharmacotherapies

7 Psychotherapies for PTSD 75
7.1 Types of Psychotherapies Used to Treat PTSD 75
7.1.1 Cognitive Therapy and Cognitive Behaviour Therapy (CBT) 75
7.1.2 Trauma-Focused CBT 75
7.1.3 Prolonged Exposure 76
7.1.4 Cognitive Processing Therapy 76
7.1.5 Eye Movement Desensitization and Reprocessing Therapy 76
7.1.6 Narrative Exposure Therapy 76
7.1.7 Stress Inoculation Training 77
7.2 Extent of Amelioration of PTSD by Psychotherapies 77
7.3 Support for Efficacy of Trauma-Focused Psychotherapies 79
7.4 Psychotherapy Amelioration of Different PTSD Symptom Clusters 80
References 81

8 Pharmacotherapies for PTSD 85
8.1 Types of Pharmacotherapies Used to Manage PTSD 85
8.1.1 Selective Serotonin Reuptake Inhibitors 85
8.1.2 Serotonin–Norepinephrine Reuptake Inhibitors 86
8.1.3 Prazosin 86
8.1.4 Benzodiazepines 86
8.1.5 Ketamine 86
8.1.6 Antipsychotics and Mood Stabilizers 88
8.1.7 3,4-Methylenedioxymethamphetamine 89
8.2 Extent of Amelioration of PTSD by Pharmacotherapies 89
8.3 Pharmacotherapy Amelioration of Different PTSD Symptom Clusters 89
References 90

9 Impact of Social Support on Veterans' PTSD Symptoms 93
9.1 The Effect of Social Support in Alleviating PTSD 93
9.2 Correlations Between Social Support and PTSD Treatment Outcomes 93
9.3 Impact of Social Support on PTSD Symptom Clusters 96
References 96

Part IV Veterans' PTSD, Grey Matter Changes and Synapse Loss

10 Cortical Grey Matter Loss Accompanying PTSD and Its Consequences . . . 101
10.1 Relationship Between Trauma, PTSD and Grey Matter Loss . . . 101
10.2 Grey Matter Loss in the Anterior Cingulate Cortex and PTSD . . . 107
10.3 Relation Between Grey Matter Loss and Resting State Networks . . . 108
10.4 Conclusions . . . 110
References . . . 111

11 Restoring Grey Matter Through Treatment . . . 113
11.1 Treatments to Ameliorate PTSD . . . 113
11.2 Treatments to Ameliorate Suicidality and Depression . . . 116
References . . . 117

12 Grey Matter Volume Changes and Synaptic Alterations . . . 121
12.1 Relationships Between Psychiatric Diseases and Changes in Dendritic Volume . . . 121
12.2 Effectiveness of Ketamine to Restore Synapses . . . 123
References . . . 126

13 Cellular and Molecular Mechanisms of Synapse Loss and Replacement in Mood and Stress Disorders: BDNF and Microglia . . . 129
13.1 Brain-Derived Neurotrophic Factor (BDNF) Molecular Mechanisms in Synapse Plasticity . . . 129
13.1.1 TrkB Receptor Activation and Downstream Effectors . . . 129
13.1.2 Activity-Dependent Versus Constitutive BDNF Release in Antidepressant Responses . . . 130
13.1.3 Integration and Long-Term Maintenance . . . 132
13.1.4 Conclusion . . . 132
13.2 Brain-Derived Neurotrophic Factor (BDNF) and Synapse Growth/Replacement . . . 132
13.2.1 Ketamine Increases BDNF and Thus Restores Synapses . . . 132
13.2.2 Repetitive Transcranial Magnetic Stimulation (rTMS) . . . 134
13.2.3 Physical Activity . . . 135
13.2.4 Conclusion . . . 136
13.3 Microglia and Synaptic Plasticity . . . 136
13.3.1 Early Evidence for Microglial Pruning . . . 136
13.3.2 Location- and Age-Dependent Mechanisms of Pruning . . . 136
13.3.3 Morphology and Phagocytic Specificity . . . 137
13.3.4 Functional Consequences and Plasticity . . . 137
13.3.5 Astrocyte–Microglia Synergy . . . 137
13.3.6 Additional Modulatory Pathways . . . 137
13.3.7 Region-Specific Refinement and Disease Implications . . . 138
13.3.8 Conclusion . . . 138

13.4 Microglial Molecular Mechanisms in Synapse Loss 138
13.4.1 Chemotactic 'Find-Me' Signals 139
13.4.2 Eat-Me Versus Don't-Eat-Me Signals 140
13.4.3 Complement-Dependent Synaptic Tagging 141
13.4.4 Fractalkine-Mediated Engulfment 141
13.4.5 Additional Opsonins and Accessory Pathways 141
13.4.6 Engulfment and Degradation 142
13.4.7 Fn14-Dependent Refinement (Experience-Dependent Phase) 142
13.4.8 Conclusion 142
References 143

Part V Ameliorating Veteran's PTSD with Neurotherapies

14 Abnormalities in Resting State Networks in PTSD and Treatment with Transcranial Magnetic Stimulation (TMS) 149
14.1 Abnormalities of Functional Activity and Connectivity of Resting State Networks in PTSD 149
14.1.1 The Default Mode Network (DMN), the Central Executive Network (CEN) and the Salience Network (SN) 149
14.1.2 Abnormalities of the DMN Associated with PTSD 152
14.2 Transcranial Magnetic Stimulation (TMS) Used to Normalize RSN Functional Activity and Connectivity 154
14.2.1 TMS Used to Treat Psychiatric Disease 154
14.2.2 Stimulating the 'PTSD Network': The Amygdala, Medial Prefrontal Cortex, and Anterolateral Temporal Lobe 155
14.2.3 Stimulation of the Dorsolateral Prefrontal Cortex to Treat PTSD 156
14.3 Other Therapies for Normalizing Abnormalities in RSNs 157
14.3.1 Trauma-Focused Cognitive Behaviour Therapy (TF-CBT) 157
14.3.2 Mindfulness-Based Interventions 157
14.3.3 Neurofeedback 158
14.3.4 Pharmacological Interventions 158
14.4 Comparison of Therapies 159
References 161

15 Abnormalities in Resting State Networks in PTSD and Treatment with Electroencephalography (EEG) and Neurofeedback Interventions 165
15.1 Electroencephalography (EEG) Methodology 165
15.2 EEG Alpha-Wave Power, PTSD and Neurofeedback 166

15.3 Increasing EEG Alpha Power in PTSD Patients Is Accompanied by Normalization of Functional Activity and Connectivity in Resting State Networks 168
15.3.1 Default Mode Network (DMN) Functional Activity 168
15.3.2 Default Mode Network (DMN) Functional Connectivity 169
15.3.3 DMN Intrinsic Functional Connectivity 169
15.3.4 DMN Extrinsic Functional Connectivity 170
15.4 fMRI Measures of Functional Activity and Connectivity Modulated by Neurofeedback 170
15.4.1 Amygdala-Centred fMRI Activity 170
15.4.2 Posterior Cingulate Cortex–Centred fMRI Activity 172
15.4.3 Comparison Between EEG-NF and fMRI-NF 172
References 173

Part VI Comparison of Interventions for Treatment of PTSD

16 Comparison of Psychotherapies, Pharmacotherapies and Neurotherapies for PTSD 177
16.1 Most Effective Psychotherapies and Pharmacotherapies 177
16.2 Comparison of Treatments for PTSD According to National and International Guidelines 178
16.3 Comparison of the Relative Effectiveness of Neuroscientific, Psychological and Pharmacological Interventions 180
16.4 Conclusion 183
References 183

Part VII The Synaptic Origins of PTSD

17 Dissociative PTSD Explained by Failure of Dendritic Integration of Synapses of Cortical Layer 5 Pyramidal Neurons . . . 189
17.1 The Central Role of Cortical Layer 5 Pyramidal Neurons (L5PNs) in Consciousness 189
17.2 Models of the Role of L5PNs in Consciousness 193
17.2.1 Dendritic Integration Theory 193
17.2.2 Integrated Information Theory Applied to Dendritic Integration Theory 195
17.3 Effects of Stress and Trauma on L5PNs 196
17.4 The Effect of Loss of L5PN Dendritic Synapses on Consciousness and PTSD 196
References 197

Part I
Vignettes

Chapter 1
Vignettes of Australian Veterans and 'Shell Shock' in World War I

1.1 Introduction

The following narrative comprises seven vignettes told by a veteran who served on the Western Front during World War I, based on entries in his diary, as well as an account of 'shell shock', the condition later called posttraumatic stress disorder (PTSD),[1] and its medical treatment at the time.

Born in 1881 in Ballarat amid the Victorian gold rush, 'Pop', as he was called by his grandson Peter, gives an account of how he thrived as a gymnastics instructor before volunteering for the Australian Imperial Force when the war erupted. The diary describes how, after arriving in France with the First Australian Division in early 1916, he participated in the Somme offensive at Pozières, where he endured relentless artillery barrages and mud-choked trenches and witnessed comrades maimed, buried alive or blown apart. He describes his memories of diving for cover from exploding shells and his fear of being mistakenly labelled a malingerer by the Royal Australian Medical Corps (RAMC), especially after realizing that he had shell shock. After Pop was relieved by the Second Division, he recounts the stories his mate Jack, who served in that Division, later described to him. Jack continued fighting through the old German lines at Pozières and observed at close-quarters bayonet combat and unimaginable carnage. Post-war, Pop struggled with nightmares, violent outbursts and strained family relationships with his daughter and wife, exacerbated by his PTSD symptoms. Pop's attempts to reintegrate into society through employment as a door-to-door salesman faltered due to his pessimistic attitude and irritable behaviour. The subsequent economic depression made it impossible for a World War I veteran on a disability pension to secure employment.

[1] The group of symptoms that now identify PTSD was clearly recognized in the seventeenth century, called 'nostalgia' by the Swiss, 'homesickness' by the Germans (Heimweb) and the French (maladie du pays) and 'to be broken' (estar roto) by the Spanish. See Gabriel (1988, chap. 4).

M. Bennett, *Veteran's Stress, Trauma and Synapse Rehabilitation*,
https://doi.org/10.1007/978-981-95-9594-5_1

Even by the time of World War II, more than 20 years after the battle for Pozières, Pop's PTSD symptoms were still evident, and the nightmares continued.

The vignettes in this chapter are presented for analysis in later chapters, which focus on the psychological ramifications of veterans' stress and trauma. In some vignettes, Pop's descriptions of his symptoms are linked to the different symptom clusters that characterize PTSD (see Tables 2.1 and 2.2), shown in square brackets.

1.2 Vignette 1: Origins and Enlistment

I was born in the Victorian town of Ballarat in 1881, where the Victorian gold rush had begun some 30 years earlier. The great prosperity this brought to the town ensured its rapid growth and enhanced both educational expansion and job opportunities. An example is the growth of physical education in the school curriculum as well as clubs devoted to such education. Gymnastics was particularly prominent and followed the curricula developed in Sweden, Germany and Britain. I showed, at an early age, particular aptitude for gymnastics, which I then specialized in as a young man in Ballarat where I subsequently became a gymnastics school teacher. Gymnastics remained the central interest of my life. But that commitment was interrupted at the age of 34 when my loyalty to the British and Australian Crown, to which I had become devoted, was tested by the opening up of hostilities of the World War I, the 'Great War' as it was called, in 1914.

I soon volunteered to join the Australian Imperial Army, which was destined to lose more men per capita than any other nation fighting in that European war, about as far away from the Antipodes as one could imagine. I arrived in France aboard a troop ship in early 1916, in time to participate in the infamous Battle of the Somme, especially at Pozières, where I witnessed the greatest level of slaughter in modern history up to that time. My experiences of that battle have had, and continue to have, a devastating effect on my life and on that of my young family since returning to Australia. Thinking I was setting off on a great adventure in 1916, I acquired a dozen threepenny-lined exercise books in which to write my impressions of the great events that I was surely going to participate in.

On arriving in England on the troop ship from Australia, I spent only a few weeks in intensive training before my division was somewhat hastily shipped to France, as the great battle for the Somme had not been going well and we were required to capture and defend the small town of Pozières (see Fig. 1.1). This was the single most dramatic event of my life, when the division was thrust directly into the thick of battle.

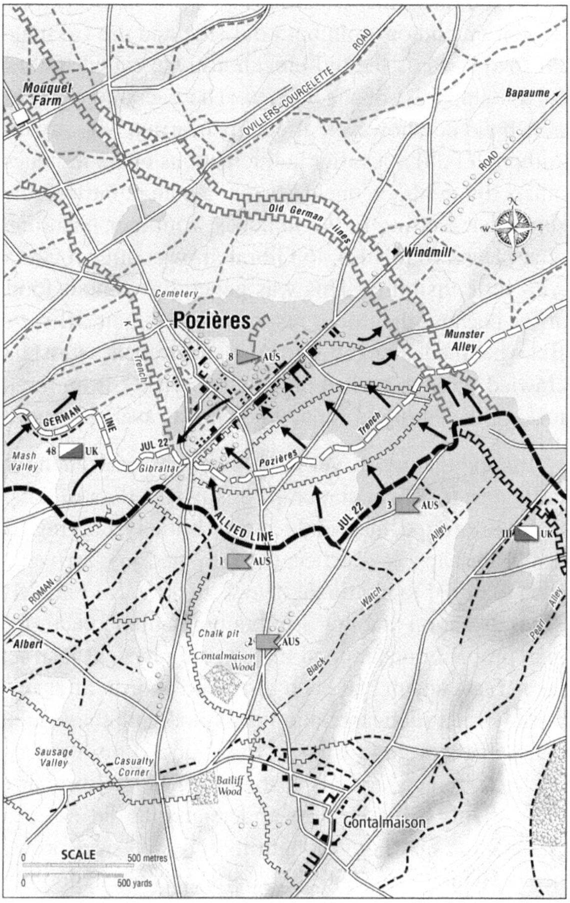

Fig. 1.1 Map of the capture of Pozières in 1916. (Map created by Laurie Whiddon, Map Illustrations)

1.3 Vignette 2: The Battle from Sausage Valley and the Chalk Pit Road to the Capture of Pozières Village

Psychological Analysis Given in Sects. 2.2, 3.4.1, 3.4.2 and 3.4.3

We were subjected to the continual sound of the bark of battery guns, the roar of field guns and the screaming sound of howitzers, all amounting to a cacophony of flying shells followed by incessant explosions as we moved up from Albert to Sausage Valley on the 23rd of July, still about 1.8 km from our objective, the capture of the village of Pozières. There was continual firing of shells from our side from behind our trenches, while at the same time the Huns[2] fired their shells at us,

[2] The 'Huns' were a nomadic tribe from the East (probably Mongolia), under the leadership of Atilla, that had invaded the Roman empire in the mid-fifth century, hence the use of this epithet by the Australians and others for the invading German troops.

sometimes at similar distances in front of the trenches. This shelling never stopped—it went on and on from both our side and the Germans' as we moved up the Chalk Pit road towards the old blockhouse we called 'Gibraltar' on the 23rd and 24th of July, south-west of the village. The cacophony of searing sounds enveloping the trench did not, however, affect my hearing, only my memory, sight, smell and taste.[3] Although I did not suffer hallucinations in the trenches from such constant shelling, others did, with, on one occasion, a man pointing his rifle at me and demanding my surrender. Getting to the trenches from rear positions, such as from the Chalk Pit along Dead Man's Rd. to Gibraltar, was often a case of crawling through the mud with little visibility. This was generally without food, as the rations had failed to arrive because the troops responsible for their delivery were likewise bogged down. But what made this effort to reach the trenches most harrowing was that the mud we crawled through reeked of a stench arising from the pulp of dead men and horses, making the eating of food, when it was available, quite repulsive.

I found that whenever a shell went over a trench I was digging, I would involuntarily dive for cover, flattening against the floor of the partly dug trench. Meanwhile, the men around me would go on blithely digging, discounting the possibility of death. Perhaps, it was because they had just arrived in the trench and had not yet been exposed to continual heavy shelling as I had. My diving for cover was a kind of reflex action and not, I felt, a premeditated act of cowardice. The newly arrived men probably understood that even a physical instructor like me could not control his nerves, which had been disturbed shortly after my arrival in that conflict. I had already seen such large numbers killed with these shells now aimed at our trenches, many of them good friends, that I hoped my bravery under fire was no longer in question by new arrivals at the trenches in Gibraltar during the terrible day of the 24th of July. This level of understanding did not extend in general to the officer class, some of whom thought that my physical prowess did not reflect my mental health, which must have been of a nervous irritability even when I volunteered in Australia. One colonel even went so far as to raise his heavy rod and bring it down with full force on my helmet, repeating the action several times while yelling, 'Pull yourself together, bloody well pull yourself together, you fool!' These actions did not prevent me from diving for cover under shell attack, an action understood by mates in the trenches.

My principal concern at the time was that my condition would be designated as a form of malingering, and that the RAMC would, in their near total ignorance of psychology and psychiatry, send me home without the possibility of a pension to support me and my young wife and my daughter aged 4 years. There was a distinct possibility that my inadvertent nervous reactions to the overhead shells would be read by RAMC officers as a class of 'unwounded shell shock', thereby excluding me from a pension. However, I noted that some British officers who had fought in the much earlier Battle of Mons (in September 1914) still displayed distressing signs of nervous strain. Indeed, one of these officers confided in me that during an

[3] Problems which also affected the civilian population.

intense shelling, he would compare his notion of time between major elevated bombardments with the reading on his watch and found that what felt like while a full day often amounted to mere seconds.

The deluge of earth following a shell explosion buried men at least up to their necks. Those in the trenches were offered the alternatives of being ripped apart by high-explosive shrapnel and, if lying down in the trenches to avoid such a fate, being buried alive in earth and debris heaved up by the shells. Such were the conditions we of the First Division endured as we advanced from the Pozières trench at Gibraltar towards our objective, Pozières village. The most intolerable strain induced by the constant exploding ordnance was not so much the risk of being killed outright by a shell, but rather being buried alive by the great masses of earth thrown up by the explosions. The strain of being entirely buried, dug out and buried again on the nerves of both officers and men was what the men came to call 'shell shock', a condition suffered by at least 20% of those classified as wounded. I gave up digging for comrades as I, like others, was overwhelmed by the hopelessness of this repetitive situation. On one occasion, as we fought our way into the village from Gibraltar and the trench, I found myself digging after an explosion to free a mate whose fingers were just protruding above the surface of the earth, searching feverishly for his face so as to dislodge dirt from his air passages and allow breathing—but this proved to be only a forlorn hope for I could not save him. Explosive ordnance was so powerful that it could remove a whole section of a trench, leaving adjacent sections relatively unaffected. One such explosion, just several yards away from my trench, killed several men instantly, while others had limbs blown off and still others were buried alive. It was amazing that I did not die. The designation 'missing in action' often referred to those missing under the dirt. If you had pooled the stock of a thousand butcher shops, it would give you a faint conception of the state of those trenches after a shelling at Pozières on the 25th of July.

I had the personal experience of frantically digging a mate out from the dirt and mud after an explosion, attempting to save him from asphyxiation. On achieving this and stepping back a few yards, I saw him blown to pieces. As we moved to consolidate our capture of Pozières village, I saw a friend, further along the trench from my position, blown in half, with one half thrown up on one side of the trench and the other on the opposite side. On other occasions, I saw men that had been buried alive by the mud thrown up by one explosion and then decapitated by another shell. Once, when I was attempting to rescue a mate completely buried in the earth following an explosion, I found while groping blindly in the earth for the buried man what appeared to be his head. On attempting to pull this free, together with his body from the enveloping soil, I found that the head was no longer attached to the body, and when the body was retrieved, both legs were found to have been blown off by the shell as well.

1.4 Vignette 3: The Battle for the Old German Lines, North–North-East of Pozières

Psychological Analysis Given in Sects. 2.2, 3.4.1 and 3.4.2

On the 26th–27th of July, the First Division was at last relieved from these horrors by men of the Second Division, whose particular objective was to take the old German lines east of the village, of which there were two (roughly parallel), designated OG1 and OG2. My best mate from the gymnastics school in Ballarat, Jack, was serving in the Second Division. In later years, when Jack and I returned home to St Kilda, amazingly not in a box, we lived not far from each other in Chapel Street and would often meet in the evening for a drink at the local Alma Hotel opposite my rented family home. There, I learnt what the Second Division had endured on taking over from the First Division, and that it involved much the same set of horrific experiences. Jack told me that conditions in the trenches, in which he hoped to survive the screaming and exploding shells, as elements of his division tried to make their way from Gibraltar to K Trench immediately west of the village, can best be described as a crowd of men crouching between blood-soaked, sand-bagged walls of the dying and the dead. These men either had been blown to pieces, making identification impossible, or had foolishly lifted their heads above the parapets and been caught by a sniper's bullet to the head. Survival in the trenches was completely serendipitous. On one occasion, a shell landed in a dugout directly in front of one of Jack's mates in K Trench but did not explode; troops in an adjacent dugout yelled out to him to shift to their part of the trench, which appeared safer as it had not recently been hit by a shell, and when he did shift, a shell fell there, blowing off his head.

Jack told me he saw many, following a shell hit, killed, buried alive or with their limbs blown off. Whole trenches could disappear under a barrage of shells, with some of the men dug out again and again. The human consequences of this were nervous breakdowns (shell shock) in both officers and men, with those going through cycles of burial and being dug out unable to continue. Nevertheless, it was essential to secure K Trench if we were to take OG1 and OG2 by fighting past the village cemetery to reach the Ovillers–Courcelette Rd., where we had a chance of taking over a track that reached along the road to cut through the OG1 and OG2 lines.

When I worked with motorized transport, I often found it came bogged down in this mud on the way to the ammunition depot behind the trenches, and there was insufficient manpower appointed to take the ammunition further. In this case, with the soldiers from India under my command, I would lead horse-drawn wagons over the shattered ground to the front line depots and even to the trenches themselves. The wagons often had to crunch their way over dead and mutilated bodies, pulverizing them further, as it was often not possible to bury the dead without adding to their numbers. On occasions I had to jump from the wagons into the morass, this charnel house, in order to guide the horses, leading to the horrific experience of walking on a pavement consisting of the remains of the dead and portions of their mutilated bodies.

On reaching the trenches as day broke, we were often presented with an appalling sight: many of the wounded were propped up among the dead bodies in horrific poses, pushed up into the parapets to rot and decay together. Collections of men, obviously killed by the same shell, lay in a cluster of grotesque figures, their bodies black and bloated in decay, with the rotting flesh peeling from their bodies. We had to push our way through this scene out of Hades to take up our positions, but what I remember most vividly was a soldier whose body had been buried up to his neck by an explosion, having his head blown off by a subsequent explosion.

Another approach to breaking the OG1 and OG2 lines was at Munster Alley on the 27th–29th of July. Jack said that, having arrived at a forward-line trench, we were then expected to attack the Huns. We left the trenches we had only just occupied, and charged across the rotting bodies of men from previous charges, to reach—if we did in the face of continuous raking machine-gun fire—the darkness of the German trenches where the enemy awaited us. Steel bayonets were used to thrust into the bodies of the Germans in the trenches, then wrenched out, thrust in again and twisted, with the rifle butt used to club and split the skulls of the enemy, spilling out their brains. But of course the Germans reciprocated, and they had the advantage in killing our men, who were exhausted from having just crawled to their own trenches and then charging over no-man's land.

We could have hundreds of thousands of men facing each other in deadly combat without seeing one man, as they were all hidden in trenches at OG1 and OG2. These were regarded by us and the Huns as citadels beyond which no one could venture without being shot by a sniper or blown to pieces by raking machine-gun fire and shells. Wounds from a single bullet fired by a sniper could be terrible: one account Jack heard from a survivor was that a bullet had entered just under his nose, striking his teeth and then proceeding along the jaw, down his throat, where it severed several arteries along the way.[4] Thousands of rounds of iron, of exploded shells, littered the ground. Jack said that he was sometimes given the task of directing Indian soldiers to clear the trenches between the explosions. On some occasions, he observed incredibly brave stretcher–bearers make trips into no-man's land under heavy machine-gun fire as well as the rain of sniper bullets, in order to rescue the wounded among the dead. On other occasions, they would be recruited to carry ammunition along the trenches to those areas that needed replenishment. How any of these stretcher–bearers survived the war was a miracle. But after constant battering and fearful losses, by the 29th of July, our Second Division took and held both OG1 and OG2 as well as Munster Alley.

[4]There were no cosmetic surgery facilities in Australia after the war for the reconstructions required for serious facial disfigurements. What could be done in the way of reconstruction was carried out at Queen Mary's Hospital in Sidcup, England, before the soldier was sent home to Australia.

1.5 Shell Shock and Its Medical Treatment in the World War I

Cases of shell shock amounted to about one third of all those discharged from the army in the World War I.[5] The average number of men lost through death or wounds suffered in the trenches in a battalion of 1000 was about 30 each month, but in some battalions, this number reached as high as 180 per month.[6] At the battle of Pozières, some 6741 Australians were killed, with many in the trenches invalided out with severe shell shock.[7] The constant rattling of the German machine guns, in some cases only a few hundred yards away, took its toll in both life and sanity, with some who could not continue blowing their heads off with their own rifles.[8]

The term 'shell shock' was first used in the medical literature by Dr. Charles Myers, a captain in the RAMC, in a 1915 article explaining his idea that shell shock was a psychological condition and should be treated as any other practical problem of 'workmen in their workplace'.[9] In contrast, others at this time took up the idea that the essence of shell shock involved a conflict between a sense of duty and all-enveloping fear.[10] However, psychology was not yet regarded as a respectable scientific subject, so the RAMC were hesitant to follow Myer's lead and resisted the idea of shell shock as a psychological condition.[11] They held that such a prognosis as psychological would be unhelpful. Most unsatisfactorily, John Collie, the leading pre-war 'expert' in 'malingering', was given the task in Britain of supervising the Houses of Recovery for those with shell shock (as well as whether those so classified would be allocated pensions). He defined 'shell shock' as a 'bankruptcy of nervous force' in a lecture entitled 'Obsession and fraud: studies in diseased personalities'.[12]

It was gradually accepted that shell shock was a psychological response to the stress of trench warfare, akin to the nervous breakdowns then recognized as the psychological response to the stress of modern civilian life. So by the end of 1915, the British Army Council decreed that shell-shock cases should have their medical reports marked with a 'W', 'if their condition was due to enemy action'—that is, the shell-shocked soldier was to be regarded as wounded and was eligible to receive a wound stripe. Alternatively, if the soldier received the designation 'S', they were to be regarded as sick and receive no such recognition. By mid-1916, the designation S was defined to include those with nervous shock not clearly related to battle, but to the general stress of battle—still not deserving of a wound stripe and certainly not

[5] Reid (2011, chap. 1).

[6] Ashworth (1980, p. 16).

[7] King (2015, p. 106).

[8] King (2015, p. 60).

[9] Myers (1915).

[10] Jones and Wessely (2005, p. 47).

[11] Reid (2011, chap. 1).

[12] Collie (1915).

a pension.[13] The essential difference between W and S designations was that the former had been in direct contact with the effects of explosives.[14] In general, the medical officers endeavoured to place those with nervous exhaustion and the psychologically wounded in the W category.[15] The S category was regarded by the troops as designating malingers, who were called 'scrimshankers'.[16] Curiously, just as the social significance of shell shock was being increasingly recognized, particularly in literary stories of the war, it fell from the parlance of the medical literature until the 1970s, and then just as an introduction to the concept of PTSD—a term which to the general public does not carry the same emotional weight as 'shell shock'.[17]

The most 'advanced' psychological medical practice for shell-shock victims involved electrical treatment, massages, strict dietary control and, in a few cases, psychotherapy.[18] The electrical treatment consisted of the application of electricity to various parts of the body, which could be a harrowing experience if the strength of the current was increased to correct what was perceived as a 'weakness of will'.[19] This technique could enter the realms of a Frankenstein horror story when strong currents were applied to the neck and throat.[20] Such faradism, as it was called, could easily be seen as a form of punishment for those suffering from shell shock, and indeed any 'success' in ameliorating the nervous condition of shell shock was almost certainly driven from fear of faradism.[21] Psychotherapy treatments of the time treated the shell-shock patient as suffering from a neurosis expressed as a consequence of the conflict between doing one's duty and trying to preserve one's life.[22]

The main medical challenges of the war were to treat multiple amputations, reconstruct obliterated facial features and return the gastrointestinal tract to proper function. These physical wounds naturally took priority over shell-shock victims, who were thought to suffer from 'hysteria', a mental disease previously associated with females. Indeed, medical officers were constantly alert to cases of malingering, in which soldiers deliberately extended a period of illness not just by exaggeration but also by enhancing their wounds or even by self-injury. If this was detected or suspected by the treating medical officer, severe discipline was brought to bear.

---

[13] Collie (1918).

[14] Jones and Wessely (2005, p. 25).

[15] Reid (2011, chap. 1).

[16] Reid (2011, chap. 1).

[17] Reid (2011, chap. 1), Jones and Wessely (2005, p. 47).

[18] Reid (2011, chap. 1).

[19] Jones and Wessely (2005, p. 33).

[20] Jones and Wessely (2005, p. 33).

[21] Reid (2011, chap. 1).

[22] Jones and Wessely (2005, p. 29).

1.6 Vignette 4: Post-war Nightmares and Violent Behaviour

Psychological Analysis Given in Sects. 2.2, 6.1 and 6.2

My nightmares were a complete replication of the traumatic events I had experienced at Pozières, which I found easy to recall as they contained a direct threat to my life.[23] The worst were the experiences seeing soldiers, my mates, being buried alive by the great masses of earth thrown up by the shells (Vignettes (v) 2 and 3). Very often a nightmare might consist of a recurring theme throughout the night of being buried, dug out and then reburied. This cycle would be repeated the following night and sometimes happened several nights in the same week [B].

The most vivid of my nightmares were those that replicated specific horrors: the friend I saw blown in half, the men surviving burial alive only to be killed by another shell and digging to find a mate's head detached from his body and then his body with the legs blown off (v2). When I woke up, these nightmarish horrors were linked to my feelings of anger, fear, grief and guilt—emotions which my family had to put up with without having the slightest idea of their origin [C, D].[24]

Very often, I was aware during the nightmare that I was dreaming and was able to reflect on the dream. But it was not much use when such lucid nightmares[25] meant I relived these experiences: crawling in mud reeking of dead men and horses as we searched for food in the trenches, the nausea that made eating anything impossible (v2) and pulverizing dead and mutilated bodies with our wagons, or with our own feet (v3).

Jack told me of his nightmares, which were not all that different from mine, and this helped forge our friendship through our shared troubles. For example, the sight of one of Jack's mates in K Trench having his head blown off by a shell after moving to a supposedly safer dugout (v3) was a recurring lucid nightmare for Jack.

On my repatriation to Australia from the battlefields in France in 1919, these vivid memories of my shell-blasted mates were ever present, colouring my daily and nightly experiences [B2]. At that time, I had become a secluded man, shunning social contact, in contrast to the very convivial person I had been before going to France, known among my fellow physical educationalists at their occasional meetings, when they were joined by their wives, as a jovial person, full of wit and funny anecdotes [D6]. These nightmares continued in the decades that passed, although their frequency and intensity became less—still present but dimmed [B1].

When the World War II began in the Pacific, as a consequence of the Japanese surprise attack on Pearl Harbor, my nightmares became exacerbated [B2]. This was particularly the case following the Japanese midget submarine attack on vessels in Sydney Harbour and subsequent shelling of Sydney coastal suburbs. Although Sydney is over 400 miles north of Melbourne, I felt that the next attack would be on

[23] Possemato et al. (2022).

[24] Simos and Berle (2023).

[25] See Miller et al. (2021).

Melbourne from midget submarines in Port Phillip Bay shelling the seaside suburb of St Kilda where we lived. I then began frantically digging a shelter in the backyard and bought an Andersen air-raid shelter to go with it. I also bought gas masks for my wife Georgina, my daughter Ava and young Peter, my grandson, a purchase prompted by my vivid memory of the gas attacks by the Germans that I had experienced on the Somme all those decades earlier [B5]. Very unfortunately, not only did the nightmares increase in frequency during this period, but so did my need to maintain constant vigilance for Japanese aircraft, which led me to frequently withdraw from any conversation for long periods, up to days at a time [E3].

I could not control outbursts of anger in relation to a perceived provocation—such as when eating dinner at the big table in the kitchen, if young Peter breached some table etiquette, like picking up his peas with a knife or failing to offer me the salt. My general feelings of alienation and guilt from killing so many people in that first war returned, with my accusing Peter on occasion that he was not worthy of the sacrifices that my mates and I had made in saving him from the Germans [E1].

My behaviour then was like that when I first returned from the Western Front all those years ago, although now, it no longer had the destructive effect on my daughter Ava, who had long grown up, married and departed our house on Chapel Street, giving birth to young Peter when she was 24. Now, however, they were being visited on her son, Peter. But when Ava was six and I had just returned from the war, she had to contend with being ignored by me, even when pleading for attention, as well as my occasional aggressive behaviour, which on one occasion led to my slapping her face. This behaviour was exacerbated by my alcoholic tendencies, which were fed by the recently built Alma Hotel opposite our house. Years later, I was mortified by my behaviour, of which I had dutifully kept an account in my threepenny exercise book diaries, now amassing dozens, from the first all those years ago before being shipped off to the Somme in 1916.

So my emotional outbursts were not just confined to nightmares but occurred in sudden violence during the daytime, shameful behaviour I could not contain [E2]. While my wife knew that the origins of this dangerous behaviour were to be found in my wartime experiences, my 5-year-old daughter was unlikely to make head or tail of them, and indeed seemed to think, according to her mother, that the outbursts were somehow due to her being a naughty girl. However, her mother insisted Ava had been a lovely girl when growing up in Melbourne during my absence at the war. I could not but wonder, as I grew older, whether my neglect of Ava and her other early experiences as a child with a shell-shocked father did not give rise to her future status as a 'good-time girl' in St Kilda during the World War II.

1.7 Vignette 5: Inability to Find Work

Psychological Analysis Given in Sect. 2.2
The small disability pension I received was not nearly sufficient to sustain my wife and children, for by then, Ava had been joined by a brother, Sam.[26] This had necessitated my finding employment, which was not easy for someone who had left school at 15 and had no qualifications other than the training in physical education decades earlier in Ballarat.

How was I to find a job in St Kilda or adjacent Prahran after being away 4 years at war? It was terribly hard. I started on the disheartening task of calling in on shopkeepers along the length of Chapel Street, looking for any kind of employment that might bring in sufficient money, which together with my war pension would pay for food, clothing and rent. The rental money was for a single-fronted semi-detached house that my family and I subsequently lived in for decades following the first war. I made inquiries at the many shops that abutted each other along the length of Chapel Street from near the beach in East St Kilda, not far from Eildon Road, to as far as the Prahran Town Hall.

Chapel Street, between High Street and Commercial Road, had the highest density of emporiums for retail shopping in Australia at that time. Magnificent buildings were to be found there at the beginning of the twentieth century. As High Street was only about a kilometre from near the corner of Inkerman Street and Chapel Street where we lived opposite the Alma Hotel, it was not a long journey to walk in search of employment among the emporiums. However, I did find the steep gradient at the beginning of this journey, before reaching Dandenong Road some 600 m away, quite physically demanding. This was particularly the case as I still suffered from a minor shrapnel wound to my left leg, and although I could have caught a cable tram,[27] that would have been of little use if I was to visit many adjacent stores between High Street and Commercial Road.

I visited more than 50 shops and departments in the large emporiums and arcades along Chapel Street, all to no avail. I was bitterly surprised that each of these would not employ me. The most frequent reason given was that Melbourne was 'glutted' with ex-servicemen, without particular qualifications, and so they had to be turned away. I was exhausted and thoroughly dejected by this experience, and began mulling over the shocking experiences of the Somme battles, and the dreaded malaise of suicidality revisited me. However, before giving up this chore, I visited by chance the monumental big store on the corner of Wattle and Chapel streets, recognizable

[26] The pension rate was nominally 50% of the average weekly wage, with more than 65% of eligible veterans receiving less than this. A peak of 283,322 war pensions were being paid in 1932. By 1938, only a year before the start of the World War II, 77,000 incapacitated soldiers and 180,000 dependants remained on pensions that by then had cost Australia nearly 148 million pounds. Their associated medical bills ran to another 8.5 million pounds. See Larsson (2004, pp. 45–59), Larsson (2009), Butler (1943, p. 960), and Parliament of Australia (1924, p. 47).

[27] Trams were not electrified until the late 1920s. See Malone (1983, p. 50).

by its horizontal appearance alleviated by vertical windows with arched heads and pediments. The signage indicated this was Foy and Gibsons (sometimes just 'Foys'), which had been erected in 1902 and was Melbourne's largest department store specializing in clothing, manchester, furnishings, furniture, food, etc. Clothing, in particular, was a highlight, as indicated by the relative floor space given over to their sales in the store, occupying several floors in what was and is still a five-storey building, with separate displays for men's clothing, shirts, bedding, furniture and hardware, and special prominence given to women's clothing. I entered the store with some trepidation, given the intimidating scale of the largest emporium I had ever seen, let alone entered. On making inquiries from a sales girl as to employment, I was directed to an office on the third floor. That required me to take a lift, which I had never experienced before, with the claustrophobic space and elevation against gravity making me feel very anxious and alert, and leading to profuse perspiration [E3]. I thought that there was no chance of someone in my visibly agitated state and somewhat down-at-heel appearance (I was wearing a rather threadbare double-breasted suite bought for my wedding in 1913) being offered employment in such an imposing organization. Much to my surprise, I was not ushered into a room for an interview, but led to a bench with a wire screen above it that allowed communication between the busy office beyond the screen and the general space of the emporium in which I now found himself. I explained to a man behind the grid that I was a returned serviceman from the Great War, and although on a pension needed employment to supplement what was a meagre amount of about 35% of the average wage.

The employment officer at Foy and Gibson said that the only positions they had available were for door-to-door salesmen, particularly those willing to sell women's clothing from the emporium's handsome catalogue, supplemented by samples of material used in such clothing. However, if I was willing to take on this job, then I would only receive a salary that was about 10% of the average salary at that time. Nevertheless, a major positive was that I would receive a commission on sales, which for a successful salesman could substantially increase his income to about 40% of the then average salary. I was then handed a Foy and Gibson catalogue that displayed both winter and summer clothes for women and was asked to sit down and peruse the catalogue in order to form an opinion as to whether I would like to try the job out. With no other employment prospects in view, I readily accepted.

1.8 Vignette 6: Inability to Remain Employed

Psychological Analysis Given in Sect. 2.2

After being given employment at Foy and Gibsons as a travelling salesman for mostly women's clothes, I found the routes I was given to canvass sales in St Kilda and Prahran quite ideal. I had lived in this area for many years, in both Blessington Street before the war and Chapel Street after the war. I then had an intuitive knowledge of the socio-economic strata in which I was attempting to make sales as I came from the same background. My sales blossomed and so did my remuneration for reasons that probably included the fact that I normally had a quiet and gentle manner. My income was now sufficient to feed my family of four, namely Georgina, Ava and Sam, as well as myself. It also allowed me to pay the rent on the single-fronted semi-detached house at 83 Chapel Street, near Inkerman Street and at the opposite end of the street from Foy and Gibson at 303 Chapel Street.

However, this situation did not last long. I continued to be deeply disturbed by noises that surprised me, such as the loud honk of a car horn near me in the street where I might be canvassing sales, noises that often reminded me of the impact of shells [E4]. On one occasion, I was displaying the current Foy and Gibson catalogue and samples of the dresses advertised to a woman in her house, when she obviously become very uncomfortable. This was because after being startled by a backfiring car passing by, I had become agitated and started to perspire profusely, and I became uncontrollably aggressive in my sales pitch [E4]. The inevitable outcome was loss of a potential sale. I found that these difficulties were exacerbated if I had experienced nightmares the night before. This was particularly the case if the nightmares involved incidents of a horrible nature that I had observed during the war, such as seeing a mate's head blown off during the battle of the Somme (v2) [B2]. Following such dreams, I felt impelled to break off all social engagements, which did not lead to appropriate conduct when attempting to interest a customer in a potential sale [D].

These difficulties inevitably led to a drop in sales, a loss in income through commissions, and finally, Foy and Gibson sacked me. These were humiliating early years after the Great War.

1.9 Vignette 7: Effect of PTSD on Family

Psychological Analysis Given in Sects. 2.2, 6.1 and 6.2

During these years, I became very depressed, blaming myself for failure to obtain employment [D3] and becoming easily irritable at the slightest provocation [E1]. What I most regret is that my beautiful blond daughter, Ava, who was only six when I returned from the Somme, attracted both my disdain on the one hand and my most violent outbursts of anger on the other [E1]. I do not know why this occurred, why I could not control myself. Perhaps, it was because Ava had a naturally effervescent personality that I could not tolerate when I was in a state of depression and emotional withdrawal [D6]. At any rate, I developed a poor relationship with Ava right from her earliest years, which did not help her own developing emotional stability. My volatility was not directed towards Sam as it was towards Ava, perhaps because he was very young and I saw in him a likeness of my early self. Sam's physicality was precocious even as a young child, and I thought he was clearly destined to follow in my early footsteps as a physical educationalist. Very unfortunately, most of my ire was directed towards Ava, with potentially the outcome that when eventually she had her own son, my grandson Peter, her emotional instability was so poor that he had a grim early childhood.[28]

Georgina would not have been able to cope with the stresses I introduced into our small house had she not learnt from the Australian Repatriation Department[29] in August 1919 that I had been hospitalized on several occasions in the United Kingdom during the latter stages of the war with shell shock. The density of family and lodgers in the semi-detached house in which we lived meant that Georgina and I had to sleep together in the one bed under the eaves in the attic. She would be awakened by my screaming in recurring nightmares of what I had observed and survived in the trenches on the Western Front (v4) [B2]. She found these as difficult to endure as my antisocial behaviour towards her and our daughter during the daylight hours [D]. She told me that it had been immediately apparent to her that I returned to Melbourne a changed man, with outbursts of anger and violence that permeated the environment of the small family, and which were in particular directed towards 6-year-old Ava [E1]. Georgina told me that it was obvious that I had returned a tormented man, with my yelling and rages greatly concerning the young family [E4]. On one occasion, I threatened Georgina that I would blow my head off in front of her with my service revolver, illegally retained after the war, and then began hitting her. When the very young Ava tried to intervene and save her mummy, I hit her also, with such force that she was knocked down [E1]. This was the worst of all my uncontrollable outbursts and the most unforgivable. On occasions, I felt forced to emotionally withdraw from Georgina and Ava, going silent for days at a time and becoming emotionally numb [D7]. Although I could not bring myself to be affectionate to Ava, Georgina at least realized what I had been through and what had brought on this terrible transition in the man she had loved and clearly desperately wanted to continue to love. Ava was never cuddled by me [D6]. It was Ava who bore the brunt of much of the psychological collapse of my character as a consequence of the war.

I had been unemployed for some years when the Great Depression began in 1929, having been sacked from Foy and Gibsons years earlier. I remained unemployed until the depression was over in 1939, and the World War II had begun. In order to feed and house the family after I was sacked, I evolved the strategy of renting out parts of our semi-detached house in Chapel Street. Such sub-letting was legal under certain provisions which I ensured we met, allowing two couples to live in the house besides my family. As our bedrooms were in the attic, this left space downstairs for two couples to each possess a bedroom, a lounge room and a small kitchen. However, only a single bathroom and an outside toilet were available to be

[28] Recounted in Bennett (2024, chap. 1).

[29] In 1918, the Australian Government created the Repatriation Commission to help returned veterans resettle in the Australian community. The commission introduced programmes to assist veterans develop their vocational skills and find jobs and supported their independence through pensions and loans. It also grants pensions and benefits and provides treatment and other services. See Department of Veterans' Affairs (n.d.).

shared between the eight people who came to occupy the house. Nevertheless, the rental from the two couples, together with my pension, was just enough to sustain the family for the years of the depression, when Ava became apprenticed as a ladies hairdresser and Sam a physical exercise instructor like his father had been in Ballarat. This reasonably stable financial situation for the family was not accompanied by an emotionally stable situation, for my outbursts, flashbacks and nightmares have never really ceased [B3].

References

Ashworth T (1980) Trench warfare 1914–1018: the live and let live system. Macmillan, London
Bennett MR (2024) Childhood stress, trauma and synapse loss. Springer, Singapore
Butler AG (1943) Official history of the Australian Army medical services in the war of 1914–1918: volume III—special problems and services. Australian War Memorial, Canberra
Collie J (1915) Obsession and fraud: studies in diseased personality. Sir John Collie's lecture. The Times, 27 January, p 5a
Collie J (1918) The management of neurasthenia and allied disorders contracted in the army. J State Med 6:11
Department of Veterans' Affairs, Australian Government (n.d.) The Repatriation Commission. https://www.dva.gov.au/about-us/who-we-are/the-repatriation-commission
Gabriel RA (1988) No more heroes: madness and psychiatry in war. Hill and Wang, New York
Jones E, Wessely S (2005) Shell shock to PTSD: military psychiatry from 1900 to the Gulf War. Psychology Press, London. https://doi.org/10.4324/9780203086827
King J (2015) The Western Front diaries: the Anzacs' own stories, battle by battle. Scribe, Brunswick
Larsson M (2004) Restoring the spirit: the rehabilitation of disabled soldiers in Australia after the Great War. Health Hist 6(2):45–59
Larsson M (2009) Who picks up the pieces? Melb Hist J 37:25–32
Malone B (1983) Chapel Street Prahran: Part One, 1834–1918. Prahran Historical Series
Miller KE, Ross RJ, Harb GC (2021) Lucid dreams in veterans with posttraumatic stress disorder include nightmares. Dreaming 31(2):117–127. https://doi.org/10.1037/drm0000163
Myers CS (1915) A contribution to the study of shell shock. Lancet 185(4772):316–320. https://doi.org/10.1016/S0140-6736(00)52916-X
Parliament of Australia (1924) Report of the Repatriation Commission, 1923–1924. Australian Government, Canberra
Possemato K, Silander N, Bellete N, Emery JB, De Stefano L, Pigeon W (2022) Characteristics of posttraumatic nightmares and their relationship to PTSD severity among combat veterans with PTSD and hazardous alcohol use. J Nerv Ment Dis 210(3):223–226. https://doi.org/10.1097/nmd.0000000000001422
Reid F (2011) Broken men: shell shock, treatment and recovery in Britain 1914–1930. Bloomsbury, London
Simos A, Berle D (2023) The phenomenology of nightmares in post-traumatic stress disorder and complex post-traumatic stress disorder. Eur J Trauma Dissociation 7(3):100335. https://doi.org/10.1016/j.ejtd.2023.100335

Part II
Veterans' PTSD Following Different Combat Experiences

Chapter 2
Clinical Identification of PTSD

2.1 DSM-5 Criteria for PTSD Symptom Clusters

Posttraumatic stress disorder (PTSD) is a mental health condition triggered by experiencing or witnessing a traumatic event. PTSD symptoms were traditionally categorized into three symptom clusters according to the fourth edition of the *Diagnostic and Statistical Manual of Mental Disorders* (DSM-4): re-experiencing, avoidance/numbing and hyperarousal.[1] The fifth edition of the manual (DSM-5) subsequently identified four main symptom clusters: intrusion, avoidance, negative alterations in cognition and mood and alterations in arousal and reactivity (Table 2.1).[2]

To be diagnosed with PTSD, an individual must meet specific criteria across these four symptom clusters, including experiencing symptoms for more than a month that cause significant distress or impairment in daily functioning. It is of prime importance to identify these symptom clusters correctly, as the choice of psychotherapy (Chap. 7) or pharmacotherapy (Chap. 8) depends upon them.

[1] American Psychiatric Association (1994).

[2] American Psychiatric Association (2013).

M. Bennett, *Veteran's Stress, Trauma and Synapse Rehabilitation*,
https://doi.org/10.1007/978-981-95-9594-5_2

Table 2.1 DSM-5 definitions of PTSD symptom clusters

Symptom cluster	Description	Criteria
Intrusion (B)	Recurrent, involuntary and intrusive distressing memories of the traumatic event(s)	B1. Recurrent, involuntary and intrusive distressing memories of the traumatic event(s)[a] B2. Recurrent distressing dreams related to the traumatic event(s)[b] B3. Dissociative reactions (e.g. flashbacks) in which the individual feels or acts as if the traumatic event(s) are re-occurring B4. Intense or prolonged psychological distress at exposure to internal or external cues that symbolize or resemble an aspect of the event(s) B5. Marked physiological reactions to reminders of the traumatic event(s)
Avoidance (C)	Persistent avoidance of stimuli associated with the traumatic event(s)	C1. Avoidance of or efforts to avoid distressing memories, thoughts or feelings about or closely associated with the traumatic event(s)[c] C2. Avoidance of or efforts to avoid external reminders (people, places, conversations, activities, objects, situations)
Negative alterations in cognition and mood (D)	Negative changes in thoughts and mood associated with the traumatic event(s)	D1. Inability to remember an important aspect of the traumatic event(s) (typically due to dissociative amnesia)[d] D2. Persistent and exaggerated negative beliefs or expectations about oneself, others or the world[d] D3. Persistent, distorted cognitions about the cause or consequences of the traumatic event(s) that lead the individual to blame themselves or others D4. Persistent negative emotional state (e.g. fear, horror, anger, guilt and shame) D5. Markedly diminished interest or participation in significant activities D6. Feelings of detachment or estrangement from others D7. Persistent inability to experience positive emotions

(continued)

Table 2.1 (continued)

Symptom cluster	Description	Criteria
Alterations in arousal and reactivity (E)	Changes in arousal and reactivity associated with the traumatic event(s)	E1. Irritable behaviour and angry outbursts (with little or no provocation), typically expressed as verbal or physical aggression towards people or objects[e] E2. Reckless or self-destructive behaviour[e] E3. Hypervigilance E4. Exaggerated startle response E5. Problems with concentration E6. Sleep disturbance (e.g. difficulty falling or staying asleep and restless sleep)

Source: Reproduced with permission from the American Psychiatric Association (2013)
DSM-5, *Diagnostic and Statistical Manual of Mental Disorders*, fifth edition; *PTSD* posttraumatic stress disorder
[a]National Institute of Mental Health (NIMH) and American Psychological Association (APA)
[b]NIMH
[c]National Center for PTSD
[d]Mayo Clinic
[e]APA

2.2 Application of the DSM-5 Criteria to the Experiences of an Australian Veteran

Chapter 1 presented the behaviour and experiences of an Australian veteran (Pop) on returning home with 'shell shock' from the battle of Pozières on the Western Front (the Somme) during World War I. The question posed here is whether Pop's symptoms as he describes them fit the criteria for PTSD according to the different symptom clusters summarized in Table 2.1. The following extracts from Chap. 1, with the vignette (v) number that contains the extract in parentheses, show there is ample descriptive evidence that Pop suffered from PTSD.

2.2.1 *Intrusion (PTSD Symptom Cluster B)*

Intrusive distressing memories: 'On my repatriation to Australia from the battlefields in France in 1919, these vivid memories of my shell-blasted mates were ever present, colouring my daily and nightly experiences'. (v4)

Distressing dreams (*nightmares*): 'Very often a nightmare might consist of a recurring theme throughout the night of being buried, dug out and then reburied. This cycle would be repeated the following night and sometimes happened several nights in the same week'. (v4) 'These nightmares continued in the decades that passed, although their frequency and intensity became less—still present but dimmed'. (v4) 'When World War II began in the Pacific, as a consequence of the

Japanese surprise attack on Pearl Harbor, my nightmares became exacerbated'. (v4) 'I found that these difficulties were exacerbated if I had experienced nightmares the night before. This was particularly the case if the nightmares involved incidents of a horrible nature that I had observed during the war, such as seeing a mate's head blown off during the battle of the Somme'. (v6) 'She would be awakened by my screaming in recurring nightmares of what I had observed and survived in the trenches on the Western Front' (v7).

Flashbacks: 'This reasonably stable financial situation for the family was not accompanied by an emotionally stable situation, for my outbursts, flashbacks and nightmares have never really ceased'. (v7)

Psychological distress: 'Although Sydney is over 400 miles north of Melbourne, I felt that the next attack would be on Melbourne from midget submarines in Port Phillip Bay shelling the seaside suburb of St Kilda where we lived. I then began frantically digging a shelter in the backyard and bought an Andersen air-raid shelter to go with it. I also bought gas masks for my wife Georgina, my daughter Ava and young Peter, my grandson, a purchase prompted by my vivid memory of the gas attacks by the Germans that I had experienced on the Somme all those decades earlier'. (v4)

Physiological responses: 'After being startled by a backfiring car passing by, I had become agitated and started to perspire profusely …'. (v6)

2.2.2 Avoidance (PTSD Symptom Cluster C)

Avoidance of distressing memories, thoughts, or feelings about the traumatic events: 'Georgina ... learnt from the Australian Repatriation Department in August 1919 that I had been hospitalized on several occasions in the United Kingdom during the latter stages of the war with shell shock'. (v7)

2.2.3 Negative Alterations in Cognition and Mood (PTSD Symptom Cluster D)

Persistent negative emotional state: 'When I woke up, these nightmarish horrors were linked to my feelings of anger, fear, grief and guilt—emotions that my family had to put up with without having the slightest idea of their origin'. (v4) 'During these years, I became very depressed, blaming myself for failure to obtain employment …'. (v7) 'Ava had a naturally effervescent personality that I could not tolerate when I was in a state of depression and emotional withdrawal'. (v7) 'On occasions, I felt forced to emotionally withdraw from Georgina and Ava, going silent for days at a time and becoming emotionally numb'. (v7). 'Ava was never cuddled by me'. (v7)

Diminished interest or participation in significant activities: 'At that time, I had become a secluded man, shunning social contact, in contrast to the very convivial person I had been before going to France, known among my fellow physical educationalists at their occasional meetings, when they were joined by their wives, as a jovial person, full of wit and funny anecdotes'. (v4) 'Following such dreams I felt impelled to break off all social engagements, which did not lead to appropriate conduct when attempting to interest a customer in a potential sale'. (v6)

2.2.4 Alterations in Arousal and Reactivity (PTSD Symptom Cluster E)

Startle responses: 'I continued to be deeply disturbed by noises that surprised me, such as the loud honk of a car horn near me in the street where I might be canvassing sales, noises that often reminded me of the impact of shells'. (v6)

Irritable behaviour: 'I could not control outbursts of anger in relation to a perceived provocation … My general feelings of alienation and guilt from killing so many people in that first war returned, with my accusing Peter on occasion that he was not worthy of the sacrifices that my mates and I had made in saving him from the Germans'. (v4) 'After being startled by a backfiring car passing by, … I became uncontrollably aggressive in my sales pitch'. (v6) 'During these years, I became … easily irritable at the slightest provocation'. (v7) 'What I most regret is that my beautiful blond daughter, Ava, who was only 6 when I returned from the Somme, attracted both my disdain on the one hand and my most violent outbursts of anger on the other'. (v7)

Aggressive behaviour: 'So my emotional outbursts were not just confined to nightmares but occurred in sudden violence during the daytime, shameful behaviour I could not contain'. (v4) 'But when Ava was 6 and I had just returned from the war, she had to contend with being ignored by me, even when pleading for attention, as well as my occasional aggressive behaviour, which on one occasion led to my slapping her face. This behaviour was exacerbated by my alcoholic tendencies …'. (v4)

'She told me that it had been immediately apparent to her that I returned to Melbourne a changed man, with outbursts of anger and violence that permeated the environment of the small family, and which were in particular directed towards 6-year-old Ava. Georgina told me that it was obvious that I had returned a tormented man, with my yelling and rages greatly concerning the young family. On one occasion, I threatened Georgina that I would blow my head off in front of her with my service revolver, illegally retained after the war, and then began hitting her. When the very young Ava tried to intervene and save her mummy, I hit her also, with such force that she was knocked down'. (v7)

Hypervigilance: 'Very unfortunately, not only did the nightmares increase in frequency during [World War II], but so did my need to maintain constant vigilance for Japanese aircraft, which led me to frequently withdraw from any conversation for long periods, up to days at a time'. (v4) 'That required me to take a lift, which I had never experienced before, with the claustrophobic space and elevation against gravity making me feel very anxious and alert, and leading to profuse perspiration'. (v5)

2.3 Measures Used to Assess PTSD Symptoms

2.3.1 The Posttraumatic Stress Disorder Checklist for DSM-5 (PCL-5)

The posttraumatic stress disorder checklist for DSM-5 (PCL-5) is a 20-item checklist used to assess self-reported symptoms against the DSM-5 criteria for PTSD. PCL-5 has been widely used in both clinical and research settings to measure PTSD symptoms and assess symptom severity. It can be administered in various forms, including the PCL-5 with Criterion A, which includes a section to identify the worst traumatic event and assess whether it meets the criteria for PTSD symptom cluster A.[3] PCL-5's psychometric properties, such as their validity in measuring what they are supposed to, are strong, reliable in that they do not fluctuate over time and responsive in being able to detect changing conditions.[4] PCL-5's structure aligns well with both the DSM-5's four-factor model and newer, proposed models of PTSD symptomatology, indicating its robustness and adaptability to different diagnostic frameworks.[5]

Table 2.2 provides detailed information on the distribution of responses to each of the 20 items in PCL-5, as well as for each of the four symptom clusters individually and in total, on a 2016 study of veterans receiving care.[6]

The means for each item in the PCL-5 item-level descriptive statistics in Table 2.2 were calculated by averaging the responses across all individuals in the sample for that particular item.[7] Total scores of between 31 and 33 have been shown to be adequate for diagnosing PTSD,[8] so the mean total score of 36.97 shows a high level of PTSD, which is not unexpected in a sample of veterans receiving care.

The steps involved in calculating an individual's PCL-5 score are:

- *Self-reporting*: The person completes the checklist by rating each of the 20 items based on their experiences over the past month, with each item rated on a five-point scale for intensity or frequency: 0, not at all; 1, a little bit; 2, moderately; 3, quite a bit; and 4, extremely.

---

[3] Weathers et al. (2013).

[4] Blevins et al. (2015).

[5] Blevins et al. (2015).

[6] Bovin et al. (2016).

[7] Bovin et al. (2016), p. 1383.

[8] Bovin et al. (2016).

Table 2.2 PTSD checklist for DSM-5 distribution responses by item, symptom cluster and in total

PCL-5 item	Item description	Mean	Standard deviation)
1 (B1)	Memories	2.08	1.29
2 (B2)	Dreams	1.69	1.40
3 (B3)	Flashbacks	1.48	1.37
4 (B4)	Cued distress	2.18	1.38
5 (B5)	Cued physical reactions	1.85	1.37
6 (C1)	Avoiding internal reminders	2.03	1.34
7 (C2)	Avoiding external reminders	2.03	1.42
8 (D1)	Dissociative amnesia	1.30	1.43
9 (D2)	Negative beliefs	1.66	1.51
10 (D3)	Blame	1.81	1.52
11 (D4)	Negative feelings	2.10	1.49
12 (D5)	Loss of interest	1.87	1.39
13 (D6)	Detachment or estrangements	2.02	1.43
14 (D7)	Numbing	1.79	1.45
15 (E1)	Irritability or aggressive behaviour	1.63	1.38
16 (E2)	Reckless behaviour	1.14	1.31
17 (E3)	Hypervigilance	2.10	1.47
18 (E4)	Exaggerated startle response	1.84	1.45
19 (E5)	Concentration	1.96	1.41
20 (E6)	Sleep	2.42	1.49
Cluster B	Intrusion	9.28	5.87
Cluster C	Avoidance	4.06	2.60
Cluster D	Negative alterations in cognition and mood	12.54	8.15
Cluster E	Alterations in arousal and reactivity	11.09	6.75
Total score	All symptoms combined	36.97	21.16

Source: Bovin et al. (2016), p. 1384, Table 3. Reproduced with permission from the American Psychological Association

- *Scoring*: The score for an item represents the intensity/frequency of that particular PTSD symptom. The item scores are added together for each symptom cluster to give a score for that cluster (e.g. in Table 2.2, the mean scores for items 1–5 total 9.28, which is the mean for cluster B). The sum of the item scores is the total score. As stated above, scores of 31 and higher indicate a diagnosis of PTSD.

2.3.2 The Clinician-Administered PTSD Scale (CAPS)

The clinician-administered PTSD scale (CAPS) is a 30-item structured interview developed for clinicians to assess PTSD symptoms according to DSM-4 and later revised for DSM-5 criteria. 'CAPS focuses on the frequency, intensity and severity ratings for individual PTSD symptoms, assessing all DSM diagnostic criteria'.[9] CAPS provides both a dichotomous diagnosis (PTSD vs non-PTSD) and a continuous measure of symptom severity.

The clinician asks the interviewee a set of questions for each PTSD symptom cluster to assess both the frequency and intensity of symptoms on a five-point scale. Frequency scores range from 0 (never), 2 (once a month), 2 (two to three times a month), 3 (once a week), to 4 (two to three times a week). Intensity scores range from 0 (none), 1 (mild), 2 (moderate), 3 (severe), to 4 (extreme). Table 2.3 shows a representative sample of the 30 questions.[10] The frequency and intensity of symptoms then provide a comprehensive evaluation of PTSD severity.

The main goal in creating CAPS was to develop an interview-determined rating scale that was comprehensive and sufficiently sound to be accepted as the measure of PTSD. To meet these criteria, CAPS scores needed to encompass individual symptoms as well as symptom clusters, allowing the user of the score to rate the severity, intensity and frequency of particular symptoms and clusters, as well as determination of not just current but also future PTSD symptom states.

There are other PTSD measurement models for symptom clusters: DSM-5, dysphoric arousal, dysphoria, externalizing behaviours, anhedonia and hybrid models. Table 2.4 presents the mapping of the CAPS-5 items against these other models.

Table 2.3 Sample interview questions from the clinician-administered PTSD scale (CAPS)

PTSD symptom cluster	Sample interview questions
Intrusion (B)	Have you ever had unwanted memories of the traumatic event(s)? (If yes) How often do these memories occur Do you ever feel as though you are reliving the traumatic event(s)? (If yes) How often do these flashbacks occur?
Avoidance (C)	Do you avoid activities, places or people that remind you of the traumatic event(s)? (If yes) How often do you engage in avoidance behaviours? Do you feel emotionally numb or detached from others? (If yes) How often do you experience emotional numbness?
Negative alterations in cognition and mood (D)	Do you have negative thoughts about yourself, others or the world? (If yes) How often do these thoughts occur? Do you feel excessive guilt or blame yourself for the traumatic event(s)? (If yes) How often do you experience feelings of guilt?
Alterations in arousal and reactivity (E)	Do you have trouble falling or staying asleep? (If yes) How often does this sleep disturbance occur? Do you experience sudden outbursts of irritability? (If yes) How often do you feel irritable?

Source: Weathers et al. (2001) with permission from the American Psychological Association

[9] Weathers et al. (2001).

[10] For the full set of CAPS questions, see Weathers et al. (2013).

Table 2.4 CAPS-5 item mapping to other PTSD measurement models

Item	Item description	DSM-5	Dysphoric arousal	Dysphoria	Ext. behaviours	Anhedonia	Hybrid
1 (B1)	Memories	INT	INT	INT	INT	INT	INT
2 (B2)	Dreams	INT	INT	INT	INT	INT	INT
3 (B3)	Flashbacks	INT	INT	INT	INT	INT	INT
4 (B4)	Cued distress	INT	INT	INT	INT	INT	INT
5 (B5)	Cued physical reactions	INT	INT	INT	INT	INT	INT
6 (C1)	Avoiding internal reminders	AVD	AVD	AVD	AVD	AVD	AVD
7 (C2	Avoiding external reminders	AVD	AVD	AVD	AVD	AVD	AVD
8 (D1)	Dissociative amnesia	NCM	NCM	DYS	NCM	NAF	NAF
9 (D2	Negative beliefs	NCM	NCM	DYS	NCM	NAF	NAF
10 (D3)	Blame	NCM	NCM	DYS	NCM	NAF	NAF
11 (D4)	Negative feelings	NCM	NCM	DYS	NCM	NAF	NAF
12 (D5)	Loss of interest	NCM	NCM	DYS	NCM	ANH	ANH
13 (D6)	Detachment or estrangement	NCM	NCM	DYS	NCM	ANH	ANH
14 (D7)	Numbing	NCM	NCM	DYS	NCM	ANH	ANH
15 (E1)	Irritability or aggressive behaviour	AAR	DAR	DYS	EXT	DAR	EXT
16 (E2)	Reckless behaviour	AAR	DAR	DYS	EXT	DAR	EXT
17 (E3)	Hypervigilance	AAR	AXA	AXA	AXA	AXA	AXA
18 (E4)	Exaggerated startle responses	AAR	AXA	AXA	AXA	AXA	AXA
19 (E5)	Concentration	AAR	DAR	DYS	DAR	DAR	DAR
20 (E6)	Sleep	AAR	DAR	DYS	DAR	DAR	DAR

Source: Weathers et al. (2018), Table 5, with permission from the American Psychological Association

AAR alterations in arousal and reactivity cluster, *ANH* anhedonia cluster, *AVD* avoidance cluster, *AXA* anxious arousal cluster, *CAPS-5* clinician-administered PTSD scale for DSM-5, *DAR* dysphoric arousal cluster, *DSM-5 Diagnostic and Statistical Manual of Mental Disorders* (fifth edn.), *DYS* dysphoria cluster, *Ext.* externalizing, *EXT* externalizing cluster, *INT* intrusions cluster, *NCM* negative alternations in cognition and mood cluster

There are high correlations between all of these different models and CAPS, emphasizing CAPS as the key assessment method for PTSD.

2.3.3 *Comparing PCL-5 and CAPS*

PCL-5 is a self-report method that is particularly suited to fast assessments of large populations. It is used for large-scale PTSD assessments. CAPS is based on interviews in clinical settings and is therefore used for individualized and small-scale assessments. It allows for more detailed analysis than PCL-5, providing an accurate assessment of the patient.

2.4 Comorbidity of PTSD and Major Depressive Disorder (MDD)

Comorbidity rates of 30–50% exist between PTSD and major depressive disorder (MDD). This results in mutual exacerbation. A study of civilian patients suffering comorbid PTSD/MDD symptoms over 12 years showed that both PTSD and MDD symptoms followed roughly the same trajectory: chronic in 24.2% of patients, gradual recovery in 14.4%, sudden relapse in 29.1% and rapid recovery in 32.2% (Fig. 2.1).

2.4.1 *MDD Symptom Overlap with DSM-5 PTSD Symptoms*

DSM-5 lists nine symptoms of MDD, most of which are the same or similar to PTSD symptoms listed in Table 2.1. The shared symptoms include those relating to negative feelings and mood, poor sleep, poor concentration and irritability (Table 2.5). Only MDD manifesting as physical symptoms—changes in appetite, weight loss or gain, fatigue and loss of energy—and suicidality are not represented among PTSD symptoms, but suicidality has been shown to be comorbid with PTSD (Chap. 4).

The psychological interplay between PTSD and MDD is influenced by factors such as guilt, loss of meaning of life and cognitive restructuring. Guilt and depression and suicidality significantly worsen PTSD severity in veterans (see Chap. 4), as does the extent of combat experience (see Chap. 3).[11] Conversely, there is an inverse relationship between PTSD and loss of meaning of life at lower levels of depression.[12] This suggests that interventions aimed at enhancing meaning of life and addressing guilt may be beneficial for individuals with comorbid PTSD and MDD.

[11] Owens et al. (2009).

[12] Owens et al. (2009).

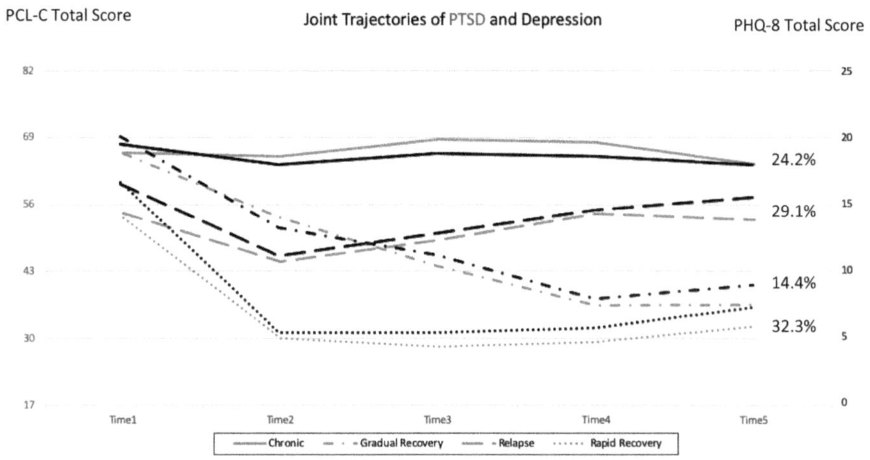

Fig. 2.1 Joint trajectories of comorbid posttraumatic stress disorder (PTSD) and major depressive disorder (MDD) adjusted for covariates in the model. Black indicates trajectories for MDD and grey indicates trajectories for PTSD. All participants screened positive for comorbid PTSD/MDD at time 1. Each time point is approximately 3 years apart. Abbreviations: PCL-C, PTSD checklist–civilian version; PHQ-8, patient health questionnaire eight-item depression scale. (Reproduced from Armenta et al. (2019). Creative Commons CC BY 4.0 (https://creativecommons.org/publicdomain/zero/1.0/))

Table 2.5 Symptoms of major depressive disorder (MDD) corresponding to PTSD symptoms

MDD symptoms	PTSD symptoms
1. Persistent sad, anxious or 'empty' mood, feelings of hopelessness	Persistent and exaggerated negative beliefs or expectations about oneself, others or the world (D2) Persistent negative emotional state (e.g. fear, horror, anger, guilt and shame) (D5) Persistent inability to experience positive emotions (D7)
2. Loss of interest or pleasure in hobbies and activities	Markedly diminished interest or participation in significant activities (D5)
3. Appetite and/or weight changes	–
4. Insomnia, early-morning awakening or oversleeping	Sleep disturbance (E6)
5. Irritability (psychomotor agitation)	Irritable behaviour and angry outbursts (with little or no provocation) (E1)
6. Physical symptoms (fatigue, loss of energy)	–
7. Feelings of guilt, worthlessness or helplessness	Persistent, distorted cognitions about the cause or consequences of the traumatic event(s) that lead the individual to blame themselves or others (D4) Persistent negative emotional state (e.g. fear, horror, anger, guilt and shame) (D5)
8. Difficulty concentrating, remembering or making decisions	Problems with concentration (E5)
9. Thoughts of death or suicide, or suicide attempts	–

Sources: Reproduced with permission of the American Psychiatric Association (APA) (2013)

Excessive rumination[13] can also exacerbate PTSD/MDD comorbidity.[14] In addition, genetic studies of twins show that about 30% of the onset of PTSD and of MDD is genetic[15] with specific genes involved in their comorbidity.[16]

2.4.2 *MDD Symptom Overlap with Dysphoria Models of PTSD*

Dysphoria refers to a state of unease or generalized dissatisfaction with life, and it captures symptoms such as emotional numbing, detachment and loss of interest in previously enjoyed activities. The four-factor dysphoria model of PTSD emphasizes the affective and mood-related symptoms of PTSD. It reveals that symptoms such as difficulty sleeping, irritability and concentration problems are common to both PTSD and MDD. As these symptoms are associated with both depression and anxiety, it seems likely that a distinct shared dysphoria/arousal construct reflecting both anxiety and depression exists.[17]

The five-factor dysphoric arousal model distinguishes 'anxious arousal' from 'dysphoric arousal', making it superior to the four-factor model for the hyperarousal symptom cluster.[18] This separation emphasizes the significant comorbid depressive-like symptoms in PTSD (e.g. E5 and E6).

Table 2.6 maps the DSM-5 PTSD symptoms from Table 2.1 to these dysphoria models as well as to other common DSM-5 PTSD models.

Standardized factor loadings represent the degree to which observed variables (i.e. symptoms) are associated with their underlying latent factors (i.e. symptom clusters) in a model. Table 2.7 shows the standardized factor loadings for the DSM-5 emotional numbing four-factor model, four-factor dysphoria model and five-factor model.

2.4.3 *Impact of Trauma Type on Comorbidity*

Different kinds of trauma, both with respect to severity and type, guide the evolution of PTSD and MDD. Exposure to different military traumas (e.g. witnessing violence and life-threatening events) gives rise to different patterns of PTSD and MDD symptoms.[19]

---

[13] Nolen-Hoeksema (2000).

[14] Michl et al. (2013).

[15] Sullivan et al. (2000).

[16] Kilpatrick et al. (2007).

[17] Elhai et al. (2011).

[18] See Armour et al. (2016) and Elhai et al. (2011).

[19] Stein et al. (2012), Table 3.

Table 2.6 Symptom mappings of the most common DSM-5 PTSD models

DSM-5 symptom	DSM5–3	DSM-5 numbing (four-factor)	Dysphoria (four-factor)[a]	Dysphoric arousal (five-factor)	Anhedonia (six-factor)	Externalizing behaviours (six-factor)	Hybrid (seven-factor)
B1: Recurrent thoughts of trauma	R	R	R	R	R	R	R
B2: Recurrent dreams of trauma	R	R	R	R	R	R	R
B3: Flashbacks	R	R	R	R	R	R	R
B4: Psychological cue reactivity	R	R	R	R	R	R	R
B5: Physiological cue reactivity	R	R	R	R	R	R	R
C1: Avoidance of thoughts of trauma	AV/ NACM	AV	AV	AV	AV	AV	AV
C2: Avoidance of reminders of trauma	AV/ NACM	AV	AV	AV	AV	AV	AV
D1: Memory impairment	AV/ NACM	NACM	D	NACM	NA	N	NA
D2: Negative beliefs	AV/ NACM	NACM	D	NACM	NA	N	NA
D3: Distorted blame	AV/ NACM	NACM	D	NACM	NA	N	NA
D4: Persistent negative emotional state	AV/ NACM	NACM	D	NACM	NA	N	NA
D5: Diminished interest in activities	AV/ NACM	NACM	D	NACM	An	N	An

(continued)

Table 2.6 (continued)

DSM-5 symptom	DSM5–3	DSM-5 numbing (four-factor)	Dysphoria (four-factor)[a]	Dysphoric arousal (five-factor)	Anhedonia (six-factor)	Externalizing behaviours (six-factor)	Hybrid (seven-factor)
D6: Feelings of detachment from others	AV/ NACM	NACM	D	NACM	An	N	An
D7: Inability to experience positive emotions	AV/ NACM	NACM	D	NACM	An	N	An
E1: Irritability or anger	A	A	D	DA	DA	EB	EB
E2: Reckless or self-destructive behaviour	A	A	A/D/D[b]	DA	DA	EB	EB
E3: Hypervigilance	A	A	A	AA	AA	AA	AA
E4: Exaggerated startle response	A	A	A/D/A[c]	AA	AA	AA	AA
E5: Concentration difficulties	A	A	D/A/D[d]	DA	DA	DA	DA
E6: Sleeping difficulties	A	A	D/A/D	DA	DA	DA	DA

*Source: Armour et al. (2016), Table 2. See also Blevins et al. (2015). With permission of the American Psychiatric Association

A alterations in arousal and reactivity, *AA* anxious arousal, *An* anhedonia, *AV* avoidance, *D* dysphoria, *DA* dysphoric arousal, *DSM-5 Diagnostic and Statistical Manual of Mental Disorders* (fifth edn.), *EB* externalizing behaviours, *N* numbing, *NA* negative affect, *NACM* negative alterations in cognitions and mood, *P* general PTSD factor, *PTSD* posttraumatic stress disorder, *R* re-experiencing

[a]Three versions of the DSM-5 dysphoria model

[b]A/D/D = anxious arousal plus two clusters for dystrophic arousal

[c]A/D/A = anxious arousal/dystrophic arousal/anxious arousal again

[d]D/A/D = dystrophic arousal as the first and third cluster, with anxious arousal in between

Table 2.7 Standardized factor loadings (four-factor, four-factor dysphoria and five-factor models) of PTSD symptoms

PTSD symptom	Four-factor model	Four-factor dysphoria model	Five-factor model
1. Intrusive thoughts	0.77	0.77	0.77
2. Nightmares	0.76	0.76	0.76
3. Reliving trauma	0.79	0.79	0.79
4. Emotional cued reactivity	0.81	0.81	0.81
5. Physiological cued reactivity	0.68	0.68	0.68
6. Avoidance of thoughts	0.80	0.81	0.81
7. Avoidance of reminders	0.87	0.87	0.87
8. Trauma-related amnesia	0.60	0.58	0.60
9. Loss of interest	0.89	0.87	0.89
10. Feeling detached	0.84	0.82	0.84
11. Feeling numb	0.82	0.79	0.82
12. Hopelessness	0.74	0.71	0.74
13. Difficulty sleeping	0.75	0.72	0.75
14. Irritable/angry	0.71	0.69	0.72
15. Difficulty concentrating	0.86	0.83	0.87
16. Hypervigilance	0.82	0.87	0.87
17. Easily startled	0.87	0.94	0.94

Source: Elhai et al. (2011), Table 2, which summarizes results from King et al. (1998) (four-factor loadings) and Simms et al. (2002) (four-factor dysphoria loadings). Reproduced with permission from Elsevier

Table 2.8 Correlations between predictors of PTSD severity

Variable	Range	Mean	SD	1[a]	2[a]	3[a]	4[a]	5[a]	6[a]
1. PTSD	17–85	48.68	21.88						
2. Age	26–89	57.08	10.62	−.26[b]					
3. Combat exposure	0–41	20.92	11.71	.27[b]	.22[b]				
4. Depression	7–28	14.53	6.53	.73[c]	−.20[b]	0.07			
5. Meaning in life	5–35	21.23	8.17	−.53[c]	0.05	−0.01	−.60[c]		
6. Guilt	25–93	64.83	16.42	.64[c]	−0.22	0.03	.63[c]	−.59[c]	

Source: Owens et al. (2009), Table 1, and Cohen (1988). Reproduced with permission from Wiley
PTSD posttraumatic stress disorder, *SD* standard deviation
[a]Strength of association: 0.1, small; 0.3, medium; and 0.5, large
[b]$p < 0.01$
[c]$p < 0.001$

A regression model shows significant predictors of PTSD severity, including age, combat exposure, depression, guilt and loss of meaning in life.[20] The correlations between these are given in Table 2.8 and help visualize how these factors interact to influence PTSD severity.

[20] Owens et al. (2009).

References

American Psychiatric Association (APA) (1994) Diagnostic and statistical manual of mental disorders, 4th edn. APA, Washington, DC

American Psychiatric Association (APA) (2013) Diagnostic and statistical manual of mental disorders, 5th edn. APA Publishing, Arlington

Armenta RF, Rush T, LeardMann CA, Millegan J, Cooper A, Hoge CW (2019) Factors associated with persistent posttraumatic stress disorder among U.S. military service members and veterans. BMC Psychiatry 19(1):1–11. https://doi.org/10.1186/s12888-018-1590-5

Armour C, Műllerová J, Elhai JD (2016) A systematic literature review of PTSD's latent structure in the Diagnostic and Statistical Manual of Mental Disorders: DSM-IV to DSM-5. Clin Psychol Rev 44:60–74. https://doi.org/10.1016/j.cpr.2015.12.003

Blevins CA, Weathers FW, Davis MT, Witte TK, Domino JL (2015) The posttraumatic stress disorder checklist for DSM-5 (PCL-5): development and initial psychometric evaluation. J Trauma Stress 28(6):489–498. https://doi.org/10.1002/jts.22059

Bovin MJ, Marx BP, Weathers FW, Gallagher MW, Rodriguez P, Schnurr PP, Keane TM (2016) Psychometric properties of the PTSD checklist for Diagnostic and Statistical Manual of Mental Disorders—Fifth Edition (PCL-5) in veterans. Psychol Assess 28(11):1379–1391. https://doi.org/10.1037/pas0000254

Cohen J (1988) Statistical power analysis for the behavioral sciences. Routledge, New York

Elhai JD, Biehn TL, Armour C, Klopper JJ, Frueh BC, Palmieri PA (2011) Evidence for a unique PTSD construct represented by PTSD's D1–D3 symptoms. J Anxiety Disord 25(3):340–345. https://doi.org/10.1016/j.janxdis.2010.10.007

Kilpatrick DG, Koenen KC, Ruggiero KJ, Acierno R, Galea S, Resnick HS, Roitzsch J, Boyle J, Gelernter J (2007) The serotonin transporter genotype and social support and moderation of posttraumatic stress disorder and depression in hurricane-exposed adults. Am J Psychiatry 164(11):1693–1699. https://doi.org/10.1176/appi.ajp.2007.06122007

King DW, Leskin GA, King LA, Weathers FW (1998) Confirmatory factor analysis of the clinician-administered PTSD SCALE: evidence for the dimensionality of posttraumatic stress disorder. Psychol Assess 10(2):90–96. https://psycnet.apa.org/doi/10.1037/1040-3590.10.2.90

Michl LC, McLaughlin KA, Shepherd K, Nolen-Hoeksema S (2013) Rumination as a mechanism linking stressful life events to symptoms of depression and anxiety: longitudinal evidence in early adolescents and adults. J Abnorm Psychol 22(2):339–352. https://doi.org/10.1037/a0031994

Nolen-Hoeksema S (2000) The role of rumination in depressive disorders and mixed anxiety/depressive symptoms. J Abnorm Psychol 109(3):504–511. https://doi.org/10.1037/0021-843X.109.3.504

Owens GP, Steger MF, Whitesell AA, Herrera CJ (2009) Posttraumatic stress disorder, guilt, depression, and meaning in life among military veterans. J Trauma Stress 22(6):654–657. https://doi.org/10.1002/jts.20460

Simms LJ, Watson D, Doebbeling BN (2002) Confirmatory factor analyses of posttraumatic stress symptoms in deployed and nondeployed veterans of the Gulf War. J Abnorm Psychol 111(4):637–647. https://doi.org/10.1037/0021-843x.111.4.637

Stein NR, Mills MA, Arditte K, Mendoza C, Borah AM, Resick PA, Litz BT (2012) A scheme for categorizing traumatic military events. Behav Modif 36(6):787–807. https://doi.org/10.1177/0145445512446945

Sullivan PF, Neale MC, Kendler KS (2000) Genetic epidemiology of major depression: review and meta-analysis. Am J Psychiatry 157(10):1552–1562. https://doi.org/10.1176/appi.ajp.157.10.1552

Weathers FW, Keane TM, Davidson JR (2001) Clinician-administered PTSD scale: a review of the first ten years of research. Depress Anxiety 13(3):132–156. https://doi.org/10.1002/da.1029

Weathers FW, Litz BT, Keane TM, Palmieri PA, Marx BP, Schnurr PP (2013) The PTSD Checklist for DSM-5 (PCL-5). National Center for PTSD, US Department of Veterans Affairs, White River Junction. https://www.ptsd.va.gov/professional/assessment/adult-sr/ptsd-checklist.asp

Weathers FW, Bovin MJ, Lee DJ, Sloan DM, Schnurr PP, Kaloupek DG, Keane TM, Marx BP (2018) The clinician-administered PTSD scale for DSM-5 (CAPS-5): development and initial psychometric evaluation in military veterans. Psychol Assess 30(3):383–395. https://doi.org/10.1037/pas0000486

Chapter 3
Combat Experience Leading to PTSD

3.1 Combat Exposure Scale

Longitudinal studies have shown the negative effects that combat exposure has on both the physical and mental well-being of those in combat,[1] with scales developed to place these effects on a quantitative basis. The combat exposure scale (CES) was originally developed to quantify the intensity of combat-related stressors experienced by military personnel in Vietnam and the likelihood that such personnel will develop posttraumatic stress disorder (PTSD).[2] A seven-item scale relating combat exposure to PTSD showed the extent to which PTSD arose in Vietnam veterans for each of the seven combat stressors:

> 1, stationed in Vietnam, 50%; 2, saw injury or death of a US serviceman, 0%; 3, fired a weapon or was fired upon in combat, 25%; 4, responsible for the death of an enemy military, 33%; 5, wounded in combat, 63%; 6, responsible for the death of an enemy civilian, 77%; 7, served a third tour of duty in Vietnam, 100%.[3]

On a scale from 1 to 40, veterans without PTSD gave a total score of 22.84 and those with PTSD a score of 29.37. Subsequent modification of this scale to move it from specification of Vietnam as the theatre of battle also used seven items as follows:

> 1, did you ever go on combat patrols; 2, were you ever under enemy fire; 3, were you ever surrounded by the enemy; 4, what percent of men in your unit were killed, wounded or missing in action; 5, how often did you fire rounds at the enemy; 6, how often did you see someone get hit by incoming or outgoing rounds; 7, how often were you in danger of being

[1] Elder Jr et al. (1997), Elder et al. (2009), Koenen et al. (2008), and Currier and Holland (2012). For an overview, see Carvalho et al. (2015).

[2] Lund et al. (1984) and Foy et al. (1984).

[3] Lund et al. (1984), p. 1325.

M. Bennett, *Veteran's Stress, Trauma and Synapse Rehabilitation*,
https://doi.org/10.1007/978-981-95-9594-5_3

injured or killed in the line of duty. This scale has been validated for its internal consistency and test–retest reliability, making it a robust tool for assessing combat-related stress.[4]

3.1.1 *Dose–Response Relationship of Combat Exposure with PTSD*

Numerous studies have demonstrated a dose–response relationship between the level of combat exposure and the incidence of PTSD. For instance, there is a strong correlation between the severity of combat exposure and the prevalence of PTSD among Vietnam veterans, with higher combat exposure significantly correlated with the likelihood of developing PTSD (Fig. 3.1).[5]

Greater exposure to combat stressors, such as direct involvement in firefights and witnessing severe injuries or deaths, heightens the risk of PTSD with correlations for specific experiences as follows: for death or injury of others, 0.39; for threat of death or injury to oneself, 0.37; for killing of others, 0.47; for participating in

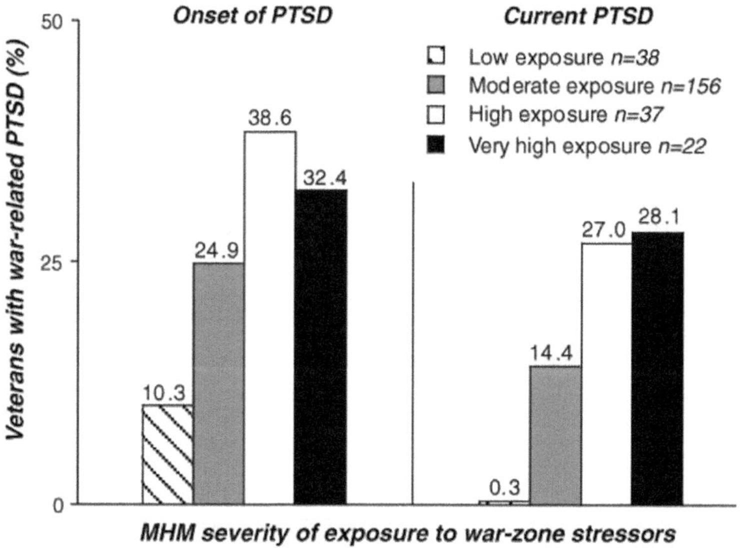

Fig. 3.1 The percentages of veterans who had suffered from war-related PTSD during their lifetime (onset of PTSD) and those who still suffered from PTSD at the time of the study (current PTSD) with increases in military historical measure (MHM) of probable severity of exposure to war-zone stressors. (Reproduced from Dohrenwend et al. (2006), Fig. 2, with permission from Science, AAAS)

[4] Keane et al. (1989), p. 54.

[5] Dohrenwend et al. (2006).

atrocities, 0.29; for harsh physical conditions, 0.36; and for insufficiency of resources, 0.45.[6] There is also no doubt that PTSD symptom severity in those occupying a non-combatant role is significantly related to war-zone stressors, such as deprivation, purposelessness and lack of unit cohesion.[7]

3.1.2 Relationship of Combat Exposure with Duration of PTSD

Thirty years after returning from the Vietnam War, 10% of veterans still had severe PTSD symptoms. Figure 3.2 is a histogram of PTSD prevalence over 14 years for veterans with low, medium and high combat exposure, showing that the prevalence does not change much over this period. Of those that served in the Vietnam War and were wounded, 40% had a definite or probable lifetime diagnosis of PTSD.[8] Indeed, 18.7% of all Vietnam veterans developed PTSD during their lifetimes as a consequence of serving.[9] Family relationships, happiness, mental health and life satisfaction were all worse among those with this continuing PTSD, although pre-existing

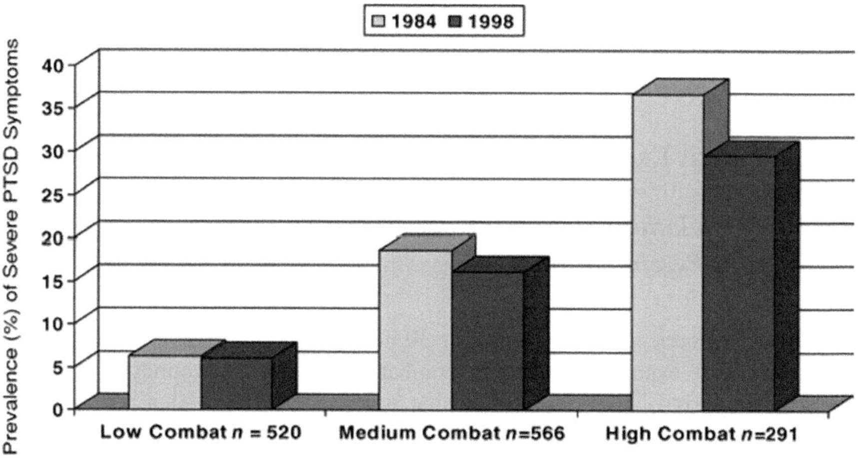

Fig. 3.2 Prevalence of severe posttraumatic stress disorder (PTSD) symptoms measured in 1984 and 1998 for Vietnam veterans (n = 1377) stratified by level of combat exposure. (Reproduced from Koenen et al. (2008), Fig. 1, with permission from Wiley)

[6] Fontana and Rosenheck (1999).

[7] McTeague et al. (2004).

[8] Pitman et al. (1989).

[9] Dohrenwend et al. (2006).

psychological conditions and social support systems moderated the impact of combat exposure on PTSD development.[10]

3.2 The Relationship Between Critical War-Zone Experiences and Mental Health Problems, Including PTSD

The type of combat exposure also plays a crucial role in determining PTSD risk. Veterans involved in more intense combat scenarios, such as direct engagements with the enemy and participating in life-threatening missions, exhibited higher rates of PTSD with correlations of 0.22 and 0.23, respectively.[11] Different types of traumatic military events, including life threats and witnessing violence, contribute to PTSD in unique ways.[12] Development of a critical war-zone experience scale consisting of seven items was used to show correlations with post-deployment mental health problems in a sample of veterans, with consistently high probabilities using three different measures of mental health problems: the depression–anxiety–stress scale, 0.90; the PTSD checklist—military version (see Sect. 2.3.1), 0.92; and the Beck depression inventory–II, 0.86 (Fig. 3.2).[13] The higher the combat exposure and to particular traumatic events, the higher the severity of PTSD,[14] with long-term studies showing that such PTSD lasts for long periods after the precipitating event(s).[15]

3.3 Combat Experience Scale

3.3.1 Correlations Between Combat Experiences and Psychopathologies

The combat experience scale comprises 30 items relating to four different categories of combat experience—violent combat, proximity to wounding or death, encountering explosive devices and combat injury (Table 3.1)[16]—that can be used to correlate experience of these experiences to the development of the psychopathological categories of PTSD, major depressive disorder (MDD), impulsivity (aggression/violence) and anxiety (Table 3.2). This scale possesses high levels of

[10] Koenen et al. (2008).

[11] Maguen et al. (2009).

[12] Stein et al. (2012).

[13] Kimbrel et al. (2014).

[14] Owens et al. (2009).

[15] Schnurr and Green (1995).

[16] Killgore et al. (2008), Watkins et al. (2016), and Osório et al. (2018).

Table 3.1 Combat experience scale categories and items as described in three different studies

Main category	Watkins et al. (2016) items	Killgore et al. (2008) items	Osório et al. (2018) items
1. Violent combat	Small arms fire	Receiving small arms fire	Coming under small arms and/or RPG fire
	Shooting	Shooting or directing fire at the enemy	Discharging a weapon in direct combat
	Attacked	Receiving incoming artillery rocket or mortar fire	Coming under mortar, artillery or rocket attack
	Clearing buildings	Clearing/searching homes or buildings	Clearing searching homes or buildings, caves or bunkers
	Encountering sniper fire	–	Encountering sniper fire
	–	Being attacked or ambushed	Experiencing a threatening situation and being unable to respond due to rules of engagement
2. Proximity to wounding or death	Handling remains	Handling human remains	Handling bodies
	–	Providing aid to wounded	Giving aid to wounded
	–	Seeing dead bodies or human remains	Seeing personnel seriously wounded or killed
	Buddy hit or shot nearby	Had a buddy shot or hit who was near you	Having a mate shot/hit who was near you
	Unable to help ill/injured	–	Seeing injured or sick women or children and being unable to help
	Knew someone killed/injured	Knowing someone seriously injured or killed	–
3. Encountering explosive devices	–	Participating in IED/mine clearing operations	–
	IED explosion	IED/booby trap exploded near you	Experiencing IED explosions
	–	Working in areas that were mined or had IEDs	Experiencing an IED
4. Combat injury	Injury	Being wounded/injured	–

Sources: Watkins et al. (2016), Table 1, Killgore et al. (2008), Table 1, and Osório et al. (2018), Table 2

IED improvised explosive device, *RPG* rocket-propelled grenade

Table 3.2 Correlations between four categories of the combat experience scale and four psychopathologies

Combat experience category	MDD	PTSD	Impulsivity	Anxiety
Violent combat[a]	*C* = 0.27[e] RR = 1.56[f] OR = 0.99[g] *C* = 0.11[h] OR = 1.29[i]	AOR = 0.51[j] *C* = 0.42[e] *C* = 0.20[k] *C* = 0.51[l] RR = 4.87[f] OR = 0.97[g] *C* = 0.32[h] OR = 7.66[i]	*C* = 0.18[e] *C* = 0.28[k] *C* = 0.07[m]	..
Proximity to wounding or death in combat[b]	OR = 2.03[g] AOR = 1.32[n]	AOR = 1.45[j] OR = 2.97[g] AOR = 4.29[n] OR = 2.47[o]	C = 0.04[m]	..
Encountering explosive devices in combat[c]	OR = 1.23[g]	AOR = 1.57[j] AOR = 1.83[p] OR = 1.20[g]	..	..
Combat injury[d]	OR = 5.7[q] OR = 5.7[r] HR = 1.47[s] OR = 1.48[t] OR = 1.40[u] *F* = 0.04[v]	OR = 3.51[o] OR = 2.10[w] OR = 3.51[o] OR = 9.10[q] OR = 9.1[r] HR = 2.92[s] OR = 2.42[t] OR = 1.35[u] *F* = 4.2[v]	..	HR = 1.34[s] OR = 1.43[t] OR = 1.27[u] *F* = 0.0[v]

.. no study for this category, *AOR* adjusted odds ratio, *C* correlation, *F* F-test, *HR* hazard ratio, *IED* improvised explosive device, *MDD* major depressive disorder, *PTSD* posttraumatic stress disorder, *RR* relative risk, *RPG* rocket-propelled grenade

Notes:

[a]Includes coming under small arms or RPG fire; discharging a weapon in direct combat; coming under mortar, artillery or rocket attack; clearing or searching homes or buildings, caves or bunkers; encountering sniper fire; experiencing a threatening situation and being unable to respond due to rules of engagement

[b]Includes experiencing hostility from civilians; handling bodies; giving aid to wounded; seeing personnel seriously wounded or killed; having a mate shot/hit who was near you; seeing injured or sick women or children and being unable to help

[c]Includes experiencing a landmine strike and experiencing an IED

[d]The higher the severity of the injury, the higher the subsequent development of PTSD (Giordano et al. 2018)

[e]Maguen et al. (2009), Table 4; $p < 0.05$

[f]Prigerson et al. (2002), Table 1

[g]Porter et al. (2018), Tables 3 and 4; $p < 0.05$ for row 2

[h]Reed-Fitzke and Lucier-Greer (2020), p. 127; $p < 0.001$

[i]Mustillo et al. (2015), Tables 3 and 4; $p < 0.001$

[j]Watkins et al. (2016), Table 1

[k]Koffel et al. (2016), Table 1; $p < 0.001$

[l]Orr et al. (1990), Table 1; $p < 0.0014$

[m]Killgore et al. (2008), Table 4; $p < 0.05$ for row 1

[n]Wells et al. (2012), Table 4, p. 90

(continued)

Table 3.2 (continued)

[o]Phillips et al. (2010), p. 1
[p]Jones et al. (2014), Table 3
[q]Giordano et al. (2018), p. 121
[r]Grieger et al. (2006), Table 2; $p < 0.0005$
[s]Walker et al. (2021), Table 2; $p < 0.001$
[t]Boscarino (1995), Table 3; $p < 0.001$
[u]Chin and Zeber (2020), Table 2; $p < 0.001$
[v]Koren et al. (2005), Table 3
[w]MacGregor et al. (2013)

Table 3.3 Correlations between combat experience and PTSD symptom clusters

Experiences in combat	Intrusions (B)	Avoidance (C)	Negative alterations in mood and cognition (D)	Alternations in arousal and reactivity (E)	Arousal (dystrophic)
Violent combat	OR = 2.43 (1.35)	OR = 1.80	OR = 1.59 (1.56)	OR = 1.33	OR = 1.90
Proximity to wounding or death in combat	OR = 3.01 (1.67)	OR = 2.10	OR = 1.59(1.05)	OR = 1.56	OR = 2.45 (1.53)
Encountering explosive devices in combat	OR = 2.14	OR = 1.68			OR = 1.97 (1.34)
Combat injury	F = 8.1	F = 1.2		F = 0.9	

Sources: OR values: Osório et al. (2018), Table 4. *F* values: Koren et al. (2005), Table 3; $p = 0.05$ for column only
F F-test, *IED* improvised explosive device, *OR* odds ratio, significant at $p < 0.001$ (adjusted OR), *PTSD* posttraumatic stress disorder, *RPG* rocket-propelled grenade

consistency[17] showing, among other things, that greater exposure to violent combat experiences, such as 'discharging a weapon in direct combat', is correlated with much greater risky experiences on returning home after the combat. Table 3.2 shows the extent to which PTSD is correlated with different combat experience categories and that, not surprisingly, combat injury is the most significant.

3.3.2 Correlations Between Combat Experiences and PTSD Symptom Clusters

Among the PTSD symptom clusters of intrusion (B), avoidance (C), negative alterations in mood and cognition (D), alterations in arousal and reactivity (E) and arousal (dystrophic) (see Chap. 2), it is intrusion—particularly re-experiencing—that is best correlated with each of the categories in the combat experience scale (see Table 3.3).

[17]Guyker et al. (2013).

3.4 Application of the Combat Experience Scale to the Experiences of an Australian Veteran

Section 2.1 applied the criteria for PTSD to the post-deployment symptoms described by an Australian World War I veteran in Chap. 1. This section extracts descriptions of the veteran's specific combat experiences described in each vignette (v) that correspond to items within three of the four categories in the combat experience scale. The category of exposure to improvised explosive devices (IEDs) is not included because IEDs were not used in that war.

As Sect. 3.3 showed very high correlations between the two categories of combat experience— violent combat and proximity to wounding or death in combat—that the veteran experienced at a high level, and the PTSD symptom cluster B (intrusion), it is therefore likely that these specific experiences were responsible for the high levels of intrusion symptoms he suffered on repatriation and much later in life.

3.4.1 Violent Combat

Coming under mortar, artillery or rocket attack: 'We were subjected to the continual sound of the bark of battery guns, the roar of field guns and the screaming sound of howitzers, all amounting to a cacophony of flying shells followed by incessant explosions … There was continual firing of shells from our side from behind our trenches, while at the same time the Germans fired their shells at us, sometimes at similar distances in front of the trenches. This shelling never stopped—it went on and on from both our side and the Germans' … Although I did not suffer hallucinations in the trenches from such constant shelling, others did, with, on one occasion, a man pointing his rifle at me and demanding my surrender'. (v2)

'I found that whenever a shell went over a trench I was digging, I would involuntarily dive for cover, flattening against the floor of the partly dug trench. Meanwhile, the men around me would go on blithely digging, discounting the possibility of death. Perhaps, it was because they had just arrived in the trench and had not yet been exposed to continual heavy shelling as I had'. (v2)

'The deluge of earth following a shell explosion buried men at least up to their necks. Those in the trenches were offered the alternatives of being ripped apart by high-explosive shrapnel and, if lying down in the trenches to avoid such a fate, being buried alive in earth and debris heaved up by the shells'. (v2)

'Explosive ordnance was so powerful that it could remove a whole section of a trench, leaving adjacent sections relatively unaffected. One such explosion, just several yards away from my trench, killed several men instantly, while others had limbs blown off and still others were buried alive'. (v2)

'Whole trenches could disappear under a barrage of shells, with some of the men dug out again and again'. (v3)

'The trenches were regarded by us and the Germans as citadels beyond which no one could venture without being … blown to pieces by raking machine gun fire and shells'. (v3)

Encountering sniper fire: 'These dead men … had foolishly lifted their heads above the parapets and been caught by a sniper's bullet to the head'. (v3) '[The trenches] were regarded by us and the Germans as citadels beyond which no one could venture without being shot by a sniper or blown to pieces by raking machine gun fire … Wounds from a single bullet fired by a sniper could be terrible …'. (v3) 'Incredibly brave stretcher–bearers [would] make trips into no-man's land under heavy machine-gun fire as well as the rain of sniper bullets, in order to rescue the wounded among the dead'. (v3)

Being in threatening situations: 'Getting to the trenches from rear positions … was often a case of crawling through the mud with little visibility. This was generally without food, as the rations had failed to arrive for the troops responsible for their delivery were likewise bogged down'. (v2) 'One such explosion, just several yards away from my trench, killed several men instantly, while others had limbs blown off and still others were buried alive. It was amazing that I did not die'. (v2)

3.4.2 Proximity to Wounding or Death

Seeing dead bodies or human remains: 'I arrived in France aboard a troop ship in early 1916, in time to participate in the infamous Battle of the Somme, especially at Pozières, where I witnessed the greatest level of slaughter in modern history up to that time'. (v1)

'If you had pooled the stock of a thousand butcher shops, it would give you a faint conception of the state of those trenches after a shelling at Pozières on the 25th of July'. (v2)

'On other occasions, I saw men that had been buried alive by the mud thrown up by one explosion and then decapitated by another shell'. (v2)

'In this case, with the soldiers from India under my command, I would lead horse-drawn wagons over the shattered ground to the front line depots and even to the trenches themselves. The wagons often had to crunch their way over dead and mutilated bodies, pulverizing them further, as it was often not possible to bury the dead without adding to their numbers. On occasions I had to jump from the wagons into the morass, this charnel house, in order to guide the horses, leading to the horrific experience of walking on a pavement consisting of the remains of the dead and portions of their mutilated bodies'. (v3)

'On reaching the trenches as day broke, we were often presented with an appalling sight: many of the wounded were propped up among the dead bodies in horrific poses, pushed up into the parapets to rot and decay together. Collections of men, obviously killed by the same shell, lay in a cluster of grotesque figures, their bodies black and bloated in decay, with the rotting flesh peeling from their bodies. We had to push our way through this scene out of Hades to take up our positions, but what I remember most vividly was a soldier whose body had been buried up to his neck by an explosion, having his head blown off by a subsequent explosion'. (v3)

Handling human remains: 'Once, when I was attempting to rescue a mate completely buried in the earth following an explosion, I found while groping blindly in the earth for the buried man what appeared to be his head. On attempting to pull this free, together with his body from the enveloping soil, I found that the head was no longer attached to the body, and when the body was retrieved, both legs were found to have been blown off by the shell as well'. (v2)

Giving aid to wounded: 'On one occasion, as we fought our way into the village from Gibraltar and the trench, I found myself digging after an explosion to free a mate whose fingers were just protruding above the surface of the earth, searching feverishly for his face so as to dislodge dirt from his air passages and allow breathing—but this proved to be only a forlorn hope for I could not save him'. (v2)

'I had the personal experience of frantically digging a mate out from the dirt and mud after an explosion, attempting to save him from asphyxiation'. (v2)

Having a mate shot/hit who was near you: 'I had the personal experience of frantically digging a mate out from the dirt and mud after an explosion, attempting to save him from asphyxiation. On achieving this and stepping back a few yards, I saw him blown to pieces. As we moved to consolidate our capture of Pozières village I saw a friend, further along the trench from my position, blown in half, with one half thrown up on one side of the trench and the other on the opposite side'. (v2)

'I had already seen such large numbers killed with these shells now aimed at our trenches, many of them good friends …'. (v2)

3.4.3 Combat Injury

Our veteran at the Somme did not receive a significant combat injury like so many around him did, except for a small shrapnel wound to the neck. However, he was hospitalized for 'shell shock', so it is those experiences that are included here.

'The strain of being entirely buried, dug out and buried again on the nerves of both officers and men was what the men came to call "shell shock", a condition suffered by at least 20% of those classified as wounded. I gave up digging for comrades as I, like others, was overwhelmed by the hopelessness of this repetitive situation'. (v2)

'My diving for cover was a kind of reflex action and not, I felt, a premeditated act of cowardice. The newly arrived men probably understood that even a physical instructor like me could not control his nerves, which had been disturbed shortly after my arrival in that conflict. I had already seen such large numbers killed with these shells now aimed at our trenches, many of them good friends, that I hoped my bravery under fire was no longer in question by new arrivals at the trenches….) This level of understanding did not extend in general to the officer class, some of whom thought that my physical prowess did not reflect my mental health, which must have been of a nervous irritability even when I volunteered in Australia. One colonel even went so far as to raise his heavy rod and bring it down with full force on my helmet, repeating the action several times while yelling, "Pull yourself together, bloody well pull yourself together, you fool!" These actions did not prevent me from diving for cover under shell attack, an action understood by mates in the trenches'. (v2)

'My principal concern at the time was that my condition would be designated as a form of malingering … There was a distinct possibility that my inadvertent nervous reactions to the overhead shells would be read by RAMC officers as a class of "unwounded shell shock", so excluding a pension'. (v2)

References

Boscarino JA (1995) Post-traumatic stress and associated disorders among Vietnam veterans: the significance of combat exposure and social support. J Trauma Stress 8(2):317–336. https://doi.org/10.1007/bf02109567

Carvalho T, Cunha M, Pinto-Gouveia J, Duarte J (2015) Portuguese version of the PTSD checklist–military version (PCL-M)–I: confirmatory factor analysis and reliability. Psychiatry Res 226(1):53–60. https://doi.org/10.1016/j.psychres.2014.11.055

Chin DL, Zeber JE (2020) Mental health outcomes among military service members after severe injury in combat and TBI. Mil Med 185(5–6):e711–e718. https://doi.org/10.1093/milmed/usz440

Currier JM, Holland JM (2012) Examining the role of combat loss among Vietnam War veterans. J Trauma Stress 25(1):102–105. https://doi.org/10.1002/jts.21655

Dohrenwend BP, Turner JB, Turse NA, Adams BG, Koenen KC, Marshall R (2006) The psychological risks of Vietnam for U.S. veterans: a revisit with new data and methods. Science 313(5789):979–982. https://doi.org/10.1126/science.1128944

Elder GH Jr, Shanahan MJ, Clipp EC (1997) Linking combat and physical health: the legacy of World War II in men's lives. Am J Psychiatry 154(3):330–336. https://doi.org/10.1176/ajp.154.3.330

Elder GH, Clipp EC, Brown JS, Martin LR, Friedman HW (2009) The life-long mortality risks of World War II experiences. Res Aging 31(4):391–412. https://doi.org/10.1177/0164027509333447

Fontana A, Rosenheck R (1999) A model of war zone stressors and posttraumatic stress disorder. J Trauma Stress 12(1):111–126. https://doi.org/10.1023/a:1024750417154

Foy DW, Sipprelle RC, Rueger DB, Carroll EM (1984) Etiology of posttraumatic stress disorder in Vietnam veterans: analysis of premilitary, military, and combat exposure influences. J Consult Clin Psychol 52(1):79–87. https://doi.org/10.1037/0022-006x.52.1.79

Giordano NA, Bader C, Richmond TS, Polomano RC (2018) Complexity of the relationships of pain, posttraumatic stress, and depression in combat-injured populations: an integrative review to inform evidence-based practice. Worldviews Evid-Based Nurs 15(2):113–126. https://doi.org/10.1111/wvn.12274

Grieger TA, Cozza SJ, Ursano RJ, Hoge C, Martinez PE, Engel CC, Wain HJ (2006) Posttraumatic stress disorder and depression in battle-injured soldiers. Am J Psychiatry 163(10):1777–1783. https://doi.org/10.1111/wvn.12274

Guyker WM, Donnelly K, Donnelly JP, Dunnam M, Warner GC, Kittleson J, Bradshaw CB, Alt M, Meier ST (2013) Dimensionality, reliability, and validity of the combat experiences scale. Mil Med 178(4):377–384. https://doi.org/10.7205/milmed-d-12-00223

Keane TM, Fairbank JA, Caddell JM, Zimering RT, Taylor KL, Mora CA (1989) Clinical evaluation of a measure to assess combat exposure. Psychol Assess 1(1):53–55. https://psycnet.apa.org/doi/10.1037/1040-3590.1.1.53

Killgore WD, Cotting DI, Thomas JL, Cox AL, McGurk D, Vo AH, Castro CA, Hoge CW (2008) Post-combat invincibility: violent combat experiences are associated with increased risk-taking propensity following deployment. J Psychiatr Res 42(13):1112–1121. https://doi.org/10.1016/j.jpsychires.2008.01.001

Kimbrel NA, Evans LD, Patel AB, Wilson LC, Meyer EC, Gulliver SB, Morissette (2014) The critical warzone experiences (CWE) scale: initial psychometric properties and association with PTSD, anxiety, and depression. Psychiatry Res 220(3):1118–1124. https://doi.org/10.1016/j.psychres.2014.08.053

Koenen KC, Stellman SD, Sommer JF, Stellman JM (2008) Persisting posttraumatic stress disorder symptoms and their relationship to functioning in Vietnam veterans: a 14-year follow-up. J Trauma Stress 21(1):49–57. https://doi.org/10.1002/jts.20304

Koffel E, Kramer MD, Arbisi PA, Erbes CR, Kaler M, Polusny MA (2016) Personality traits and combat exposure as predictors of psychopathology over time. Psychol Med 46(1):209–220. https://doi.org/10.1017/S0033291715001798

Koren D, Norman D, Cohen A, Berman J, Klein EM (2005) Increased PTSD risk with combat-related injury: a matched comparison study of injured and uninjured soldiers experiencing the same combat events. Am J Psychiatry 162(2):276–282. https://doi.org/10.1176/appi.ajp.162.2.276

Lund M, Foy D, Sipprelle C, Strachan A (1984) The combat exposure scale: a systematic assessment of trauma in the Vietnam war. J Clin Psychol 40(6):1323–1328. https://doi.org/10.1002/1097-4679(198411)40:6<1323::aid-jclp2270400607>3.0.co;2-i

MacGregor AJ, Tang JJ, Dougherty AL, Galarneau MR (2013) Deployment-related injury and posttraumatic stress disorder in US military personnel. Injury 44(11):1458–1464. https://doi.org/10.1016/j.injury.2012.10.009

Maguen S, Metzler TJ, Litz BT, Seal KH, Knight SJ, Marmar CR (2009) The impact of killing in combat on mental health symptoms and related functioning. J Trauma Stress 22(5):435–443. https://doi.org/10.1002/jts.20451

McTeague LM, McNally RJ, Litz BT (2004) Prewar, war-zone, and postwar predictors of posttraumatic stress in female Vietnam veteran health care providers. Mil Psychol 16(2):99–114. https://psycnet.apa.org/doi/10.1207/S15327876MP1602_2

Mustillo SA, Kysar-Moon A, Douglas SR, Hargraves R, Wadsworth SM, Fraine M, Frazer NL (2015) Overview of depression, post-traumatic stress disorder, and alcohol misuse among active duty service members returning from Iraq and Afghanistan, self-report and diagnosis. Mil Med 180(4):419–427. https://doi.org/10.7205/milmed-d-14-00335

Orr SP, Claiborn JM, Altman B, Forgue DF, de Jong JB, Pitman RK, Herz LR (1990) Psychometric profile of posttraumatic stress disorder, anxious, and healthy Vietnam veterans: correlations with psychophysiologic responses. J Consult Clin Psychol 58(3):329–335. https://doi.org/10.1037/0022-006x.58.3.329

Osório C, Jones N, Jones E, Robbins I, Wessely S, Greenberg N (2018) Combat experiences and their relationship to post-traumatic stress disorder symptom clusters in UK military personnel deployed to Afghanistan. Behav Med 44(2):131–140. https://doi.org/10.1080/08964289.2017.1288606

Owens GP, Steger MF, Whitesell AA, Herrera CJ (2009) Posttraumatic stress disorder, guilt, depression, and meaning in life among military veterans. J Trauma Stress 22(6):654–657. https://doi.org/10.1002/jts.20460

Phillips CJ, Leardmann CA, Gumbs GR, Smith B (2010) Risk factors for posttraumatic stress disorder among deployed US male marines. BMC Psychiatry 10:52. https://doi.org/10.1186/1471-244x-10-52

Pitman RK, Altman B, Macklin ML (1989) Prevalence of posttraumatic stress disorder in wounded Vietnam veterans. Am J Psychiatry 146(5):667–669. https://doi.org/10.1176/ajp.146.5.667

Porter B, Hoge CW, Tobin LE, Donoho CJ, Castro CA, Luxton DD, Faix D (2018) Measuring aggregated and specific combat exposures: associations between combat exposure measures and posttraumatic stress disorder, depression, and alcohol-related problems. J Trauma Stress 31(2):296–306. https://doi.org/10.1002/jts.22273

Prigerson HG, Maciejewski PK, Rosenheck RA (2002) Population attributable fractions of psychiatric disorders and behavioral outcomes associated with combat exposure among US men. Am J Public Health 92(1):59–63. https://doi.org/10.2105/ajph.92.1.59

Reed-Fitzke K, Lucier-Greer M (2020) The buffering effect of relationships on combat exposure, military performance, and mental health of U.S. military soldiers: a vantage point for CFTs. J Marital Fam Ther 46(2):321–336. https://doi.org/10.1111/jmft.12402

Schnurr PP, Green BL (1995) Trauma and health: physical health consequences of exposure to extreme stress. American Psychological Association, Washington, DC

Stein NR, Mills MA, Arditte K, Mendoza C, Borah AM, Resick PA, Litz BT (2012) A scheme for categorizing traumatic military events. Behav Modif 36(6):787–807. https://doi.org/10.1177/0145445512446945

Walker LE, Watrous J, Poltavskiy E, Howard JT, Janak JC, Pettey WBP, Zarzabal LA, Sim A, Gundlapalli A, Stewart IJ (2021) Longitudinal mental health outcomes of combat-injured service members. Brain Behav 11(5):e02088. https://doi.org/10.1002/brb3.2088

Watkins K, Sudom K, Zamorski M (2016) Association of combat experiences with post-traumatic stress disorder among Canadian military personnel deployed in support of the mission in Afghanistan. Mil Behav Health 4(3):285–292. https://doi.org/10.1080/21635781.2016.1153538

Wells TS, LeardMann CA, Fortuna SO, Smith B, Smith TC, Ryan MA, Boyko EJ, Blazer D, Millennium Cohort Study Team (2012) A prospective study of depression following combat deployment in support of the wars in Iraq and Afghanistan. Am J Public Health 100(1):90–99. https://doi.org/10.2105/ajph.2008.155432

Chapter 4
Suicidality Comorbid with PTSD

4.1 PTSD and Suicidality: An Overview

Meta-analyses show that highly significant correlations exist between suicide ideation or plans (suicidality) and posttraumatic stress disorder (PTSD),[1] with an odds ratio (OR) of about 3.0.[2] As shown in Chap. 2, PTSD is characterized by symptom clusters of intrusion, such as re-experiencing the traumatic event(s) (criteria B); avoidance of thinking about or reminders of the traumatic event(s) (criteria C), negative alterations in cognition and mood (criteria D), and alterations in arousal and reactivity (hyperarousal) (criteria E). Table 4.1 shows that of these symptom clusters, E (hyperarousal) shows the most consistently high correlation with suicidality, followed by D (negative alterations in cognition and mood).[3] The hyperarousal cluster includes symptoms, such as insomnia, irritability, hypervigilance and an exaggerated startle response. Such symptoms can exacerbate feelings of distress and impulsivity, leading to higher suicide risk. Hyperarousal symptoms have been shown to contribute to suicide risk, beyond the effects of other PTSD symptom clusters and comorbid depression.[4]

Hyperarousal is also correlated with a greater risk of premature death among veterans who do not necessarily reach the symptom threshold for PTSD. This occurs as a consequence of hyperarousal alone enhancing non-suicidal high-risk behaviours (HRBs), such as driving recklessly, taking health risks (e.g., excessive smoking and drinking alcohol) and carrying a weapon. Figure 4.1 shows the comparatively high correlation between hyperarousal and HRBs. Given the magnitude of hyperarousal's impact, clinical assessment protocols should prioritize identification and

[1] Panagioti et al. (2012), Nock et al. (2009), and Pompili et al. (2013).

[2] For a recent meta-analysis, see Gómez and López (2022).

[3] Panagioti et al. (2017).

[4] Panagioti et al. (2012).

M. Bennett, *Veteran's Stress, Trauma and Synapse Rehabilitation*,
https://doi.org/10.1007/978-981-95-9594-5_4

Table 4.1 Summary of PTSD symptom clusters correlated with suicidality and with suicide or attempted suicide

Outcome	Intrusion (B)	Avoidance (C)	Negative alterations in cognition and mood (D)	Alterations in arousal and reactivity (E)
Suicidality (suicide ideation)	*C* = 0.34[a] *C* = 0.21[b] OR = 1.12[d] *C* = 0.27[e] IRR = 1.052[f] OR = 0.99[i] *C* = 0.32[h] *C* = 0.60[l] Beta = 0.08[j] Beta = −0.115[n] *C* = 0.34[k] Beta = 0.155[p] Chi^2 = 2.52[m] OR = 2.05[q] *C* = 0.329[o]	*C* = 0.32[a] *C* = 0.19[b] OR = 1.13[d] *C* = 0.24[e] IRR = 1.00[f] OR = 0.93[i] *C* = 0.26[h] *C* = 0.67[l] Beta = −0.15[j] Beta = 0.016[n] *C* = 0.23[k] Beta = 0.095[p] Chi^2 = 2.33[m] OR = 2.49[q] *C* = −0.105[o]	*C* = 0.47[a] AOR = 1.57[c] *C* = 0.42[b] IRR = 1.063[f] *C* = 0.39[h] Beta = 0.16[j] *C* = 0.38[k] Chi^2 = 7.62[m] *C* = 0.104[o]	*C* = 0.43[a] AOR = 2.33[c] *C* = 0.32[b] *C* = 0.42[g] OR = 1.01[d] *C* = 0.33[e] IRR = 0.990[f] OR = 1.28[i] *C* = 0.36[h] *C* = 0.67[l] Beta = 0.182[n] *C* = 0.27[k] Beta = 0.250[p] Chi^2 = 9.52[m] OR = 1.10[q]
Suicide attempt or suicide	AOR = 1.154[f] OR = 1.34[s] Chi^2 = 1.55[m] OR = 1.58[t]	AOR = 1.118[f] OR = 1.27[s] Chi^2 = 5.36[m]	AOR = 1.126[f] Chi^2 = 4.48[m]	*C* = 0.1[r] AOR = 0.865[f] OR = 1.14[s] Chi^2 = 3.97[m] OR = 0.72[t]

AOR adjusted odds ratio, *beta* beta hypothesis test, *C* correlations, with $p < 0.01$, *chi*^{2} chi-squared test, *IRR* incidence rate ratio, *OR* odds ratio

Notes:

[a]Stanley et al. (2021), Fig. 3
[b]Hellmuth et al. (2012), Table 2, p. 530
[c]Selaman et al. (2014), Table 2; $p = 0.05$
[d]Barr et al. (2016), Table 2, p. 469
[e]Ben-Ya'acov and Amir (2004)
[f]Boffa et al. (2017), Tables 3 and 4
[g]Kwon et al. (2021), p. 6
[h]Brown et al. (2018), Table 4, p. 242
[i]Briere et al. (2015), Table 2
[j]Pietrzak et al. (2015), Table 2, p. 524
[k]Guerra and Calhoun (2011), Table 3, p. 16
[l]Panagioti et al. (2017), Table 2, p. 649
[m]Legarreta et al. (2015), Table 3, p. 282
[n]Surís et al. (2011), Table 2, p. 607
[o]Bell and Nye (2007), Table II, p. 1145
[p]Steyn et al. (2013), Table 1, p. 9
[q]Lemaire and Graham (2011), Table 4, p. 237
[r]Stanley et al. (2019), Fig. 2
[s]Selaman et al. (2014), Table 3, p. 250
[t]Watkins et al. (2017), Table 4, p. 12

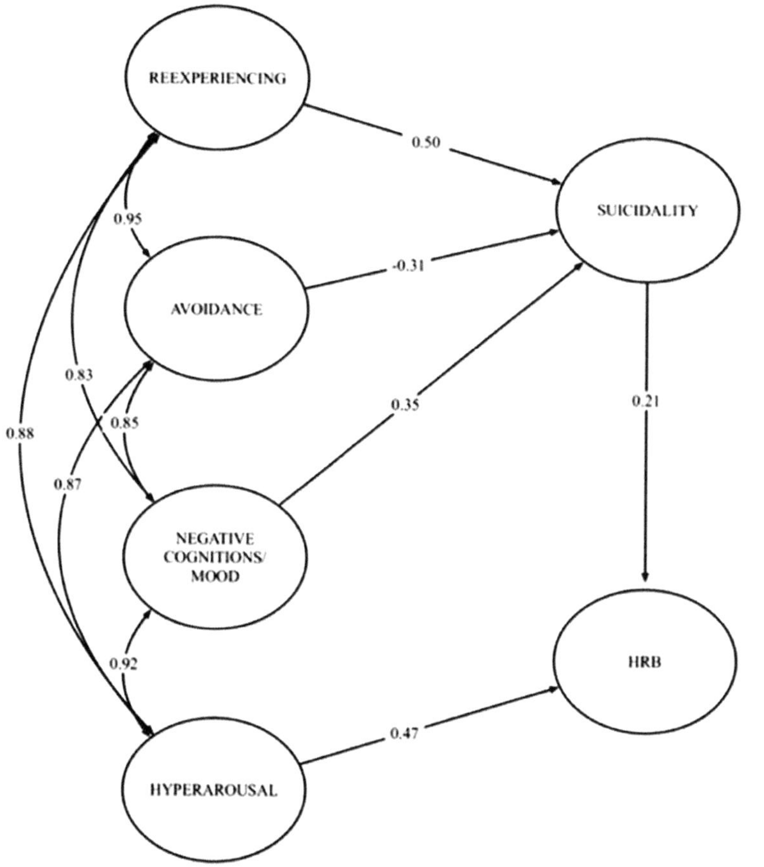

Fig. 4.1 Factor structure and significant direct effects between PTSD symptom clusters, suicidality and high-risk behaviour (HRB) latent factors. Unidirectional arrows indicate regression relationships; bidirectional arrows indicate correlational relationships. (Reproduced from Barr et al. (2018), Fig. 2, with permission from the American Psychological Association)

early management of these symptoms to interrupt trajectories towards suicidal behaviour.

4.2 The Impact of Deployment-Related Trauma

Deployment-related traumatic events, such as exposure to violent combat, witnessing wounding or death in combat and encountering explosive devices in combat, are significant risk factors for suicidality among military personnel. A meta-analysis found that 'exposure to killing and atrocities' leads 'to a 43% increased risk for suicide-related outcomes'.[5] Increased exposure to these events heightens the risk for suicide ideation, plans and attempts (Table 4.2).

Table 4.2 shows that the experience of violent combat nearly doubles the odds of suicidal ideation and quintuples the odds of suicide attempts. Proximity to death confers a 55% increase in ideation risk and a more than twofold increase in attempts. Encounters with explosives (such as improved exploding devices or landmines)

Table 4.2 Correlations between different types of combat experience and suicidality or attempted suicide

Outcome	Violent combat[a]	Proximity to wounding or death[b]	Encountering explosive devices[c]
Suicidality	AOR = 2.05[d] OR = 1.26[e]	AOR = 1.55[d] AOR = 1.17[f] OR = 1.99[g] OR = 1.4[h] OR = 1.98[i] OR = 1.3[h] OR = 1.35[e]	–
Suicide attempt	AOR = 5.32[d] HR = 1.55[j] OR = 1.56[e]	AOR = 2.25[d] $C = 0.03$[k] $C = 0.09$[k] $C = 0.14$[l] $C = 0.14$[l] HR = 1.34[j] OR = 1.46[e]	HR = 1.83[j]

AOR adjusted odds ratio (adjusted for age, sex, race, marital status, education, income, military force type and military rank), *C* Pearson correlation, *HR* hazard ratio, *IED* improvised explosive device, *OR* odds ratio, *RPG* rocket-propelled grenade

Notes:

[a]Includes coming under small arms or RPG fire; discharging a weapon in direct combat; coming under mortar, artillery or rocket attack; clearing or searching homes or buildings, caves or bunkers; encountering sniper fire; experiencing a threatening situation and being unable to respond due to rules of engagement

[b]Includes experiencing hostility from civilians; handling bodies; giving aid to wounded; seeing personnel seriously wounded or killed; having a mate shot/hit nearby; seeing injured or sick women or children and being unable to help; the combined effect here for suicide-related outcomes is $C = 0.12$, corresponding to a 43% increased risk (Bryan et al. 2015)

[c]Includes experiencing a landmine strike; experiencing an IED (Osório et al. 2018, Table 2)

[d]Sareen et al. (2017), Tables 2, 3, and 4

[e]Nichter et al. (2020), Tables 4 and 5

[f]Sareen et al. (2007), Table 3

[g]Maguen et al. (2012)

[h]Thoresen and Mehlum (2008)

[i]Sareen et al. (2007)

[j]LeardMann et al. (2021), Table 3

[k]Griffith and Vaitkus (2013)

[l]Fontana et al. (1992)

[5] Bryan et al. (2015).

yields an 83% increase in attempts. Collectively, these findings underscore the necessity of screening for specific combat exposures during post-deployment evaluations and integrating trauma-informed interventions to address both PTSD and emerging suicidality.

4.3 Impact of Psychosocial Factors

4.3.1 Impulsivity

Among psychiatric inpatients with PTSD, those with higher levels of impulsivity are more likely to be admitted for suicide-related concerns,[6] suggesting that impulsivity exacerbates the risk of suicidal behaviour in individuals with PTSD.

4.3.2 Transition to Civilian Life

The post-discharge period is dangerous due to a number of stress factors, such as loss of military structure and support, mental health challenges, potential substance abuse and financial or employment difficulties. These can lead to feelings of isolation, depression and anxiety, all of which are significant risk factors for suicide.[7]

4.3.3 Mental Health Issues

After returning from Iraq to the United States, 35% of war veterans in their first year sought mental health support related to combat experiences, leading to PTSD comorbid with major depressive disorder (MDD) and concomitant suicidality.[8]

4.3.4 Lack of Social Support

In Canada, suicidality among veterans is highly associated with depression and anxiety.[9] Only about 30% of veterans in the United States with depression and anxiety access mental health support in a given year. The failure of the remaining 70%

[6] Stanley et al. (2021).

[7] Castro et al. (2015).

[8] Hoge et al. (2006).

[9] Thompson et al. (2014a, b).

to obtain support to ameliorate their depression and anxiety puts them at high risk for suicide.[10] For example, in California, about 50% of returning veterans with mental or physical problems do not receive care, and at least 10% of these veterans suffer from suicidality.[11]

4.3.5 Substance Abuse

US veterans of the Iraq and Afghanistan wars often turned to alcohol or drugs in order to cope with mental health issues. Of those suffering from either or both PTSD and MDD, between 55% and 75% are 4.5 times more likely to be diagnosed with substance-use disorders.[12] Such disorders increase impulsivity and thereby exacerbate the risk of suicide.

4.4 Conclusion

PTSD constitutes a potent risk factor for suicidality among military and veteran populations. Hyperarousal symptoms are the most consistent predictors of suicidal ideation, attempts and HRBs. Combat exposure, impulsivity, post-discharge transition challenges, comorbid psychiatric conditions, social support deficits and substance abuse all interact synergistically to elevate suicide risk.

References

Barr NU, Sullivan K, Kintzle S, Castro CA (2016) PTSD symptoms, suicidality and non-suicidal risk to life behavior in a mixed sample of pre-and post-9/11 veterans. Soc Work Ment Health 14(5):465–473. https://doi.org/10.1080/15332985.2015.1081666

Barr N, Kintzle S, Sullivan K, Castro C (2018) Suicidality and nonsuicidal high-risk behavior in military veterans: how does PTSD symptom presentation relate to behavioral risk? Traumatology 24(1):55–61. https://psycnet.apa.org/doi/10.1037/trm0000133

Bell JB, Nye EC (2007) Specific symptoms predict suicidal ideation in Vietnam combat veterans with chronic post-traumatic stress disorder. Mil Med 172(11):1144–1147. https://doi.org/10.7205/milmed.172.11.1144

Ben-Ya'acov Y, Amir M (2004) Posttraumatic symptoms and suicide risk. Pers Individ Dif 36(6):1257–1264. https://doi.org/10.1016/s0191-8869(03)00003-5

Boffa JW, Stanley IH, Hom MA, Norr AM, Joiner TE, Schmidt NB (2017) PTSD symptoms and suicidal thoughts and behaviors among firefighters. J Psychiatr Res 84:277–283. https://doi.org/10.1016/j.jpsychires.2016.10.014

[10] Hom et al. (2017).

[11] Castro et al. (2015).

[12] Seal et al. (2011).

Briere J, Godbout N, Dias C (2015) Cumulative trauma, hyperarousal, and suicidality in the general population: a path analysis. J Trauma Dissociation 16(2):153–169. https://doi.org/10.1080/15299732.2014.970265

Brown LA, Contractor A, Benhamou K (2018) Posttraumatic stress disorder clusters and suicidal ideation. Psychiatry Res 270:238–245. https://doi.org/10.1016/j.psychres.2018.09.030

Bryan CJ, Griffith JE, Pace BT, Hinkson K, Bryan AO, Clemans TA, Imel ZE (2015) Combat exposure and risk for suicidal thoughts and behaviors among military personnel and veterans: a systematic review and meta-analysis. Suicide Life Threat Behav 45(5):633–649. https://doi.org/10.1111/sltb.12163

Castro CA, Kintzle S, Hassan AM (2015) The state of the American veteran: the Los Angeles County veterans study. USC School of Social Work, Los Angeles

Fontana A, Rosenheck R, Brett E (1992) War zone traumas and posttraumatic stress disorder symptomatology. J Nerv Ment Dis 180(12):748–755. https://doi.org/10.1097/00005053-199212000-00002

Gómez EG, López MC (2022) Revisión sistemática de la relación entre sintomatología postraumática y conducta suicida. Rev Psicopatol Psicol Clín 27(1):73–90. https://doi.org/10.5944/rppc.27757

Griffith J, Vaitkus M (2013) Perspectives on suicide in the army national guard. Armed Forces Soc 39:628–653. https://doi.org/10.1177/0095327X12471333

Guerra VS, Calhoun PS (2011) Examining the relation between posttraumatic stress disorder and suicidal ideation in an OEF/OIF veteran sample. J Anxiety Disord 25(1):12–18. https://doi.org/10.1016/j.janxdis.2010.06.025

Hellmuth JC, Stappenbeck CA, Hoerster KD, Jakupcak M (2012) Modeling PTSD symptom clusters, alcohol misuse, anger, and depression as they relate to aggression and suicidality in returning U.S. veterans. J Trauma Stress 25(5):527–534. https://doi.org/10.1002/jts.21732

Hoge CW, Auchterlonie JL, Milliken CS (2006) Mental health problems, use of mental health services, and attrition from military service after returning from deployment to Iraq or Afghanistan. JAMA 295(9):1023–1032. https://doi.org/10.1001/jama.295.9.1023

Hom MA, Stanley IH, Schneider ME, Joiner TE (2017) A systematic review of help-seeking and mental health service utilization among military service members. Clin Psychol Rev 53:59–78. https://doi.org/10.1016/j.cpr.2017.01.008

Kwon A, Lee HS, Lee SH (2021) The mediation effect of hyperarousal symptoms on the relationship between childhood physical abuse and suicidal ideation of patients with PTSD. Front Psych 12:613735. https://doi.org/10.3389/fpsyt.2021.613735

LeardMann CA, Matsuno RK, Boyko EJ, Powell TM, Reger MA, Hoge CW, Millennium Cohort Study (2021) Association of combat experiences with suicide attempts among active-duty US service members. JAMA Netw Open 4(2):e2036065. https://doi.org/10.3389/fpsyt.2021.613735

Legarreta M, Graham J, North L, Bueler CE, McGlade E, Yurgelun-Todd D (2015) DSM-5 posttraumatic stress disorder symptoms associated with suicide behaviors in veterans. Psychol Trauma 7(3):277–285. https://doi.org/10.1037/tra0000026

Lemaire CM, Graham DP (2011) Factors associated with suicidal ideation in OEF/OIF veterans. J Affect Disord 130(1–2):231–238. https://doi.org/10.1016/j.jad.2010.10.021

Maguen S, Metzler TJ, Bosch J, Marmar CR, Knight SJ, Neylan TC (2012) Killing in combat may be independently associated with suicidal ideation. Depress Anxiety 29(11):918–923. https://doi.org/10.1002/da.21954

Nichter B, Hill M, Norman S, Haller M, Pietrzak RH (2020) Impact of specific combat experiences on suicidal ideation and suicide attempt in U.S. military veterans: results from the National Health and Resilience in Veterans Study. J Psychiatr Res 130:231–239. https://doi.org/10.1016/j.jpsychires.2020.07.041

Nock MK, Hwang I, Sampson N, Kessler RC, Angermeyer M, Beautrais A, Borges G, Bromet E, Bruffaerts R, de Girolamo G, de Graaf R, Florescu S, Gureje O, Haro JM, Hu C, Huang Y, Karam EG, Kawakami N, Kovess V, Levinson D, Posada-Villa J, Sagar R, Tomov T, Viana MC, Williams DR (2009) Cross-national analysis of the associations among mental disorders

and suicidal behavior: findings from the WHO World Mental Health Surveys. PLoS Med 6(8):e1000123. https://doi.org/10.1371/journal.pmed.1000123
Osório C, Jones N, Jones E, Robbins I, Wessely S, Greenberg N (2018) Combat experiences and their relationship to post-traumatic stress disorder symptom clusters in UK military personnel deployed to Afghanistan. Behav Med 44(2):131–140. https://doi.org/10.1080/08964289.2017.1288606
Panagioti M, Gooding PA, Tarrier N (2012) A meta-analysis of the association between post-traumatic stress disorder and suicidality: the role of comorbid depression. Compr Psychiatry 53(7):915–930. https://doi.org/10.1016/j.comppsych.2012.02.009
Panagioti M, Angelakis I, Tarrier N, Gooding P (2017) A prospective investigation of the impact of distinct posttraumatic (PTSD) symptom clusters on suicidal ideation. Cognit Ther Res 41(4):645–653. https://doi.org/10.1007/s10608-016-9829-2
Pietrzak RH, Tsai J, Armour C, Mota N, Harpaz-Rotem I, Southwick SM (2015) Functional significance of a novel 7-factor model of DSM-5 PTSD symptoms: results from the National Health and Resilience in Veterans study. J Affect Disord 174:522–526. https://doi.org/10.1016/j.jad.2014.12.007
Pompili M, Sher L, Serafini G, Forte A, Innamorati M, Dominici G, Lester D, Amore M, Girardi P (2013) Posttraumatic stress disorder and suicide risk among veterans: a literature review. J Nerv Ment Dis 201(9):802–812. https://doi.org/10.1097/nmd.0b013e3182a21458
Sareen J, Cox BJ, Afifi TO, Stein MB, Belik SL, Meadows G, Asmundson GJ (2007) Combat and peacekeeping operations in relation to prevalence of mental disorders and perceived need for mental health care: findings from a large representative sample of military personnel. Arch Gen Psychiatry 64(7):843–852. https://doi.org/10.1001/archpsyc.64.7.843
Sareen J, Afifi TO, Taillieu T, Cheung K, Turner S, Stein MB, Zamorski MA (2017) Deployment-related traumatic events and suicidal behaviors in a nationally representative sample of Canadian Armed Forces personnel. Can J Psychiatry 62(11):795–804. https://doi.org/10.1177/0706743717699174
Seal KH, Cohen G, Waldrop A, Cohen BE, Maguen S, Ren L (2011) Substance use disorders in Iraq and Afghanistan veterans in VA healthcare, 2001–2010: implications for screening, diagnosis, and treatment. Drug Alcohol Depend 116(1–3):93–101. https://doi.org/10.1016/j.drugalcdep.2010.11.027
Selaman ZM, Chartrand HK, Bolton JM, Sareen J (2014) Which symptoms of post-traumatic stress disorder are associated with suicide attempts? J Anxiety Disord 28(2):246–251. https://doi.org/10.1016/j.janxdis.2013.12.005
Stanley IH, Hom MA, Boffa JW, Joiner TE (2019) PTSD symptom clusters and suicide attempts among high-risk military service members: a three-month prospective investigation. J Consult Clin Psychol 87(1):67–78. https://doi.org/10.1037/ccp0000350
Stanley IH, Marx BP, Keane TM, Vujanovic AA (2021) PTSD symptoms among trauma-exposed adults admitted to inpatient psychiatry for suicide-related concerns. J Psychiatr Res 133:60–66. https://doi.org/10.1016/j.jpsychires.2020.12.001
Steyn R, Vawda N, Wyatt GE, Williams JK, Madu SN (2013) Posttraumatic stress disorder diagnostic criteria and suicidal ideation in a South African Police sample. Afr J Psychiatry 16(1):19–22. https://doi.org/10.4314/ajpsy.v16i1.3
Surís A, Link-Malcolm J, North CS (2011) Predictors of suicidal ideation in veterans with PTSD related to military sexual trauma. J Trauma Stress 24(5):605–608. https://doi.org/10.1002/jts.20674
Thompson JM, Van Til L, Sareen J, Pietrzak RH, Zamorski MA (2014a) Suicidal ideation and behavior in Canadian Armed Forces personnel: deployment, mental disorders, and suicidal behaviors. Can J Psychiatr 59(2):64–72
Thompson JM, Zamorski MA, Sweet J, VanTil L, Sareen J, Pietrzak RH, Hopman WH, MacLean MB, Pedlar D (2014b) Roles of physical and mental health in suicidal ideation in Canadian Armed Forces Regular Force veterans. Can J Public Health 105(2):e109–e115. https://doi.org/10.17269/cjph.105.4217

Thoresen S, Mehlum L (2008) Traumatic stress and suicidal ideation in Norwegian male peacekeepers. J Nerv Ment Dis 196(11):814–821. https://doi.org/10.1097/nmd.0b013e31818b6a92
Watkins LE, Sippel LM, Pietrzak RH, Hoff R, Harpaz-Rotem I (2017) Co-occurring aggression and suicide attempt among veterans entering residential treatment for PTSD: the role of PTSD symptom clusters and alcohol misuse. J Psychiatr Res 87:8–14. https://doi.org/10.1016/j.jpsychires.2016.12.009

Chapter 5
Criminality Comorbid with PTSD

5.1 Veteran PTSD and Criminality

In 1995, the lifetime prevalence of PTSD in the general population of the United States was 7.8%.[1] About 26% of US veterans returning from the Iraq and Afghanistan wars were subsequently involved in physical aggression,[2] in some cases extending to severe violence (9%).[3] Of those that presented themselves for medical support, 41% had PTSD[4] and showed high levels of anger and hostility,[5] which extended over a lifetime.[6] Some aggressive behaviour was correlated with high levels of PTSD numbing and hyperarousal symptoms,[7] leading to intimate partner violence,[8] and likely to involve them being arrested.[9]

---

[1] Kessler et al. (1995).

[2] Elbogen et al. (2014).

[3] Elbogen et al. (2014).

[4] Sayer et al. (2010).

[5] Jakupcak et al. (2007).

[6] Flanagan et al. (2014).

[7] Flanagan et al. (2014).

[8] Taft et al. (2011).

[9] Elbogen et al. (2012), table.

M. Bennett, *Veteran's Stress, Trauma and Synapse Rehabilitation*,
https://doi.org/10.1007/978-981-95-9594-5_5

5.2 Alcoholism as an Amplifying Risk Factor

The proportion of veterans with PTSD symptoms and comorbid alcohol misuse that commit serious violence is high (35.9%) compared with those that have PTSD without the comorbid condition (10%). The rates of serious violence are still lower in veterans with neither PTSD nor alcohol misuse (5.3%).[10]

5.3 Exposure to Combat as an Amplifying Risk Factor

Engagement in combat leads to PTSD symptoms (see Chap. 3), which have a direct effect on aggression among veterans.[11] This leads to violent offending on returning from combat and gives rise to a criminal record (hazard ratio of 1.53 for combat versus noncombat roles (6.3% vs. 2.4%). Such violent offending is associated with the hyperarousal cluster of PTSD.[12] The greater the number or intensity of traumatic combat missions, the greater the extent of violent offending.[13]

5.4 Synthesis of Findings

The synthesis of these findings in Table 5.1 reveals a complex relationship between PTSD, its symptom clusters, comorbid alcohol misuse, combat exposure and various forms of criminal behaviour among veterans. Key patterns emerge: there is a more than threefold increase in serious violence among veterans with comorbid PTSD and alcohol misuse; and the likelihood of serious violence is proportional to combat mission intensity.

---

[10] Elbogen et al. (2014), Seal et al. (2011), and Murphy and Turgoose (2019).

[11] Taft et al. (2007), fig. z, Flanagan et al. (2014), and Begić and Jokić-Begić (2001).

[12] Macmanus et al. (2013).

[13] Macmanus et al. (2013), Gallaway et al. (2012), and Kessler et al. (1995).

Table 5.1 Correlations of different forms of criminality with PTSD in veterans from the Iraq and Afghanistan wars

Form of criminality	Correlations with PTSD in veterans
Violent offences (e.g. assault and homicide)	OR = 0.91[a] *C* = 2.5[b] OR = 4.1[c] *C* = 0.27[d]
Substance abuse-related crimes (e.g. DUI and drug possession)	OR = 3.18[e]
Domestic violence (e.g. intimate partner violence)	OR = 1.63[a] OR = 4.6[f]
Public disturbances (e.g. disorderly conduct)	OR = 3.18[g]

C correlations, *DUI* driving under the influence, *OR* odds ratio, *PTSD* posttraumatic stress disorder
Notes:
[a]Elbogen et al. (2014), Table 3, p. 371
[b]Jakupcak et al. (2007), Table 1, p. 949
[c]Stappenbeck et al. (2014), Table 1, p. 15
[d]Gallaway et al. (2012), Table V, p. 364
[e]Seal et al. (2011), Table 3, p. 97
[f]Sayer et al. (2010), Table 5, p. 595
[g]Begić and Jokić-Begić (2001), Table II, p. 672

References

Begić D, Jokić-Begić N (2001) Aggressive behavior in combat veterans with post-traumatic stress disorder. Mil Med 166(8):671–676. https://pubmed.ncbi.nlm.nih.gov/11515314/

Elbogen EB, Johnson SC, Newton VM, Straits-Troster K, Vasterling JJ, Wagner HR, Beckham JC (2012) Criminal justice involvement, trauma, and negative affect in Iraq and Afghanistan war era veterans. J Consult Clin Psychol 80(6):1097–1102. https://doi.org/10.1037/a0029967

Elbogen EB, Johnson SC, Wagner HR, Sullivan C, Taft CT, Beckham JC (2014) Violent behavior and post-traumatic stress disorder in US Iraq and Afghanistan veterans. Br J Psychiatry 197(3):231–238. https://doi.org/10.1192/bjp.bp.113.134627

Flanagan JC, Teer A, Beylotte FM, Killeen TK, Back SE (2014) Correlates of recent and lifetime aggression among veterans with co-occurring PTSD and substance use disorders. Ment Health Subst Use 7(4):315–328. https://doi.org/10.1080/17523281.2014.924986

Gallaway MS, Fink DS, Millikan AM, Bell MR (2012) Factors associated with physical aggression among US Army soldiers. Aggress Behav 38(5):357–367. https://doi.org/10.1002/ab.21436

Jakupcak M, Conybeare D, Phelps L, Hunt S, Holmes HA, Felker B, Klevens M, McFall ME (2007) Anger, hostility, and aggression among Iraq and Afghanistan War veterans reporting PTSD and subthreshold PTSD. J Trauma Stress 20(6):945–954. https://doi.org/10.1002/jts.20258

Kessler RC, Sonnega A, Bromet E, Hughes M, Nelson CB (1995) Posttraumatic stress disorder in the National Comorbidity Survey. Arch Gen Psychiatry 62(6):593–602. https://doi.org/10.1001/archpsyc.1995.03950240066012

Macmanus D, Dean K, Al Bakir M, Iversen AC, Hull L, Fahy T, Wessely S, Fear NT (2013) Violent offending by UK military personnel deployed to Iraq and Afghanistan: a data linkage cohort study. Lancet 381(9870):907–917. https://doi.org/10.1016/s0140-6736(13)60354-2

Murphy D, Turgoose D (2019) Exploring patterns of alcohol misuse in treatment-seeking UK veterans: a cross-sectional study. Addict Behav 92:14–19. https://doi.org/10.1016/j.addbeh.2018.11.044

Sayer NA, Noorbaloochi S, Frazier P, Carlson KF, Gravely A, Murdoch M (2010) Reintegration problems and treatment interests among Iraq and Afghanistan combat veterans receiving VA medical care. Psychiatr Serv 61(6):589–597. https://doi.org/10.1176/ps.2010.61.6.589

Seal KH, Cohen G, Waldrop A, Cohen BE, Maguen S, Ren L (2011) Substance use disorders in Iraq and Afghanistan veterans in VA healthcare, 2001–2010: implications for screening, diagnosis, and treatment. Drug Alcohol Depend 116(1–3):93–101. https://doi.org/10.1016/j.drugalcdep.2010.11.027

Stappenbeck CA, Hellmuth JC, Simpson T, Jakupcak M (2014) The effects of alcohol problems, PTSD, and combat exposure on nonphysical and physical aggression among Iraq and Afghanistan war veterans. Psychol Trauma 6(1):65–72. https://doi.org/10.1037/a0031468

Taft CT, Street AE, Marshall AD, Dowdall DJ, Riggs DS (2007) Posttraumatic stress disorder, anger, and partner abuse among Vietnam combat veterans. J Fam Psychol 21(2):270–277. https://doi.org/10.1037/0893-3200.21.2.270

Taft CT, Watkins LE, Stafford J, Street AE, Monson CM (2011) Posttraumatic stress disorder and intimate relationship problems: a meta-analysis. J Consult Clin Psychol 79(1):22–33. https://doi.org/10.1037/a0022196

Chapter 6
Family Impact of a Veteran's PTSD

6.1 Children

6.1.1 PTSD Symptoms of a Parent and Abuse of Child

PTSD symptoms are correlated with poorer parenting[1] and violent behaviour[2] involving the abuse of children,[3] with the extent of child abuse increasing with the severity of PTSD.[4] Pop's behaviour towards his daughter Ava and grandson Sam, given in Chap. 1, is an example of such abuse. Those with PTSD emotionally neglect their children, resulting in poor psychological development.[5] The various PTSD symptom clusters (see Chap. 2) of a parent correlate to different forms of abuse of their children as follows (Table 6.1):

- Re-experiencing, intrusions (B): Parents exhibit intrusive memories and flashbacks that can distract from attentive caregiving. This cluster is positively correlated with child emotional abuse and neglect.
- Avoidance, associated stimuli (C): Efforts to avoid trauma reminders reduce parental responsiveness, linked to moderate associations with emotional neglect and smaller but significant beta coefficients for emotional abuse.
- Numbing, negative alterations in mood and cognition (D): Perhaps the most deleterious for parent–child bonds, numbing symptoms correlate with physical abuse.

[1] Sullivan et al. (2016), Gewirtz et al. (2010), and Van Ee et al. (2016).

[2] Jordan et al. (1992).

[3] Jakupčević and Ajduković (2011), Table 2.

[4] Fredman et al. (2019).

[5] Van Ee et al. (2016).

M. Bennett, *Veteran's Stress, Trauma and Synapse Rehabilitation*,
https://doi.org/10.1007/978-981-95-9594-5_6

Table 6.1 Correlations of PTSD symptom clusters with different forms of child abuse

Form of childhood abuse	PTSD	Re-experiencing, intrusions (B)	Avoidance, associated stimuli (C)	Numbing, negative mood and cognition (D)	Arousal, changes in reactivity (E)
Child abuse potential	*A* = 0.51[a] OR = 1.019[b]	–	–	–	–
Physical abuse	OR = 1.936[d], Ratio, 4.0[c]	–	–	*C* = 0.15[e]	–
Emotional abuse	–	*A* = 0.03[f]	*A* = 0.02[f]	–	*A* = 0.03[f]
Emotional neglect	*C* = 0.25[g]	*A* = 0.09[f] *C* = 0.15[g]	*A* = 0.08[f] *C* = 0.23[g]	–	*C* = 0.19[g] *A* = 0.07[f]

Source: Van Ee et al. (2016)
A (beta) association, *C* correlation, *OR* odds ratio, *PTSD* posttraumatic stress disorder, *R* ratio of PTSD parent to control parent
Notes:
[a]Fredman et al. (2019), Fig. 1; $p < 0.001$
[b]Sullivan et al. (2016), Table 2; $p < 0.05$
[c]Jakupčević and Ajduković (2011), Table 1; $p < 0.05$
[d]Leen-Feldner et al. (2011), Table 2; $p < 0.05$
[e]Lauterbach et al. (2007), Table 1; $p < 0.01$ (physical abuse here includes extreme parent–child aggression)
[f]Van Ee et al. (2016), Table 2; $p < 0.05$ for row 5, except column 4
[g]Samper et al. (2004), Table 2; $p < 0.01$

- Arousal, changes in reactivity (E): Hyperarousal symptoms, such as irritability and startle responses, show moderate correlations with emotional neglect, indicating that heightened reactivity can escalate into punitive or neglectful parenting.

6.1.2 *Consequential Effects on Child*

Children of PTSD-affected parents receive limited emotional support, which undermines their psychological development. Longitudinal studies demonstrate that these children exhibit higher rates of externalizing behaviours, such as aggression, as well as internalizing symptoms, like anxiety and depression. The elevated child abuse potential scores among PTSD parents further predict adverse outcomes in school performance and peer relationships.

6.2 Partners

PTSD has been associated with physical and psychological aggression of intimate partners.[6] As with children, the PTSD symptoms of emotional numbing leads to a deterioration of relationships with a partner.[7] Table 6.2 presents the quantitative observations relating PTSD symptoms of a parent to different forms of negative behaviour with an intimate partner, as follows:

- Overall relationship quality: Relationship satisfaction correlates negatively with avoidance and numbing.
- Aggression and violence: Re-experiencing symptoms are linked to aggressive behaviours, and arousal clusters contribute further, demonstrating that trauma-related irritability and intrusive distress can spill over into partner maltreatment.
- Disagreements (discord): Both re-experiencing and avoidance symptom clusters show positive correlations with disagreements. Some regression analyses report beta coefficients indicating that numbing predicts increases in conflict.
- Lack of intimacy (contact): Emotional numbing is inversely associated with physical and emotional intimacy, pointing to significant relational withdrawal among PTSD-affected individuals.

[6] Taft et al. (2011), Yambo and Johnson (2014), Sullivan et al. (2016), and Caselli and Motta (1995)
[7] Possemato et al. (2015a, b); see also Cook et al. (2004) and Birkley et al. (2016).

Table 6.2 Correlations of PTSD symptom clusters with different forms of maltreatment of a partner

Relationship behaviour	PTSD	Re-experiencing, intrusions (B)	Avoidance, associated stimuli (C)	Numbing, negative alterations in mood and cognition (D)	Arousal, changes in reactivity (E)	Arousal (dystrophic)
Overall relationship quality	–	$C = -0.15$[a] $C = -0.01$[b] $Z = 0.17$[d]	$C = -0.27$[a] $C = -0.13$[b] $Z = 0.28$[d]	$C = -0.30$[b, c] $C = -0.27$[a] $Z = 0.32$[d]	$C = -0.11$[b] $Z = 0.25$[d]	$C = -0.19$[a]
Aggression and violence	$C = 0.44$[e]	$C = 0.21$[a] $C = -0.03$[b] $Z = 0.10$[d]	$C = 0.21$[a] $C = -0.03$[b] $Z = 0.10$[d]	$C = 0.17$[a] $C = 0.11$[b, c] $Z = 0.18$[d]	$C = 0.30$[a] $C = 0.05$[b] $Z = 0.20$[d]	$C = 0.27$[a]
Disagreements (discord)	$C = 0.36$[e] $C = 0.60$[f]	$\beta = -0.145$[g] $C = 0.06$[b] $Z = 0.21$[d]	$\beta = 0.024$[g] $C = 0.00$[b] $Z = 0.30$[d]	$\beta = 0.327$[g] $C = 0.29$[b, c] $Z = 0.47$[d]	$\beta = -0.006$[g] $C = 0.18$[b] $Q = 0.29$[d]	–
Lack of intimacy (contact)	–	$C = -0.35$[h] $\beta = -0.217$[g] $C = -0.05$[b] $Z = 0.25$[d]	$C = -0.36$[h] $\beta = 0.034$[g] $C = -0.11$[b] $Z = 0.42$[d]	$C = -0.49$[h] $\beta = 0.33$[g] $\beta = -0.33$[b, c] $\beta = -0.26$[d] $C = -0.28$[b, c] $Z = 0.35$[d]	$C = -0.39$[h] $\beta = 0.022$[g] $C = -0.17$[b] $Z = 0.25$[d]	–

β beta coefficient, *C* correlation, *PTSD* posttraumatic stress disorder, *Z* association (effect size)

Notes:

[a]Rodrigues (2014), Table 2; $p < 0.01$

[b]Erbes (2011) in Wadsworth and Riggs (2011), Table 3.3

[c]$p < 0.05$ for Wadsworth and Riggs (2011), Table 3.3, rows 1 and 3, $p < 0.10$ for row 4

[d]Birkley et al. (2016), Table 2; $p < 0.001$ for nearly all values

[e]Taft et al. (2011), Tables 1 and 2 (a meta-analysis)

[f]MacDonald et al. (1999), Table 1; $p < 0.001$

[g]Possemato et al. (2015a, b)

[h]Cook et al. (2004), Table 3; $p < 0.001$

6.3 Implications

The findings clearly show that there is a need to target specific PTSD symptom clusters through appropriate interventions in order to mitigate family violence and promote healthier parent–child and partner dynamics.

References

Birkley EL, Eckhardt CI, Dykstra RE (2016) Posttraumatic stress disorder symptoms, intimate partner violence, and relationship functioning: a meta-analytic review. J Trauma Stress 29(5):397–405. https://doi.org/10.1002/jts.22129

Caselli LT, Motta RW (1995) The effect of PTSD and combat level on Vietnam veterans' perceptions of child behavior and marital adjustment. J Clin Psychol 51(1):4–12. https://doi.org/10.1002/1097-4679(199501)51:1<4::AID-JCLP2270510102>3.0.CO;2-E

Cook JM, Riggs DS, Thompson R, Coyne JC, Sheikh JI (2004) Posttraumatic stress disorder and current relationship functioning among World War II ex-prisoners of war. J Fam Psychol 18(1):36–45. https://doi.org/10.1037/0893-3200.18.1.36

Erbes CR (2011) Couple functioning and PTSD in returning OIF soldiers: preliminary findings from the readiness and resilience in National Guard soldiers project. In: MacDermid Wadsworth S, Riggs D (eds) Risk and resilience in U.S. military families. Springer, pp 47–62

Fredman SJ, Le Y, Marshall AD, Garcia Hernandez W, Feinberg ME, Ammerman RT (2019) Parents' PTSD symptoms and child abuse potential during the perinatal period: direct associations and mediation via relationship conflict. Child Abuse Negl 90:66–75. https://doi.org/10.1016/j.chiabu.2019.01.024

Gewirtz AH, Polusny MA, DeGarmo DS, Khaylis A, Erbes CR (2010) Posttraumatic stress symptoms among National Guard soldiers deployed to Iraq: associations with parenting behaviors and couple adjustment. J Consult Clin Psychol 78(5):599–610. https://doi.org/10.1037/a0020571

Jakupčević KK, Ajduković M (2011) Risk factors of child physical abuse by parents with mixed anxiety–depressive disorder or posttraumatic stress disorder. Croat Med J 52(1):25–34. https://doi.org/10.3325/cmj.2011.52.25

Jordan BK, Marmar CR, Fairbank JA, Schlenger WE, Kulka RA, Hough RL, Weiss DS (1992) Problems in families of male Vietnam veterans with posttraumatic stress disorder. J Consult Clin Psychol 60(6):916–926. https://doi.org/10.1037/0022-006X.60.6.916

Lauterbach D, Bak C, Reiland S, Mason S, Lute MR, Earls L (2007) Quality of parental relationships among persons with a lifetime history of posttraumatic stress disorder. J Trauma Stress 20(2):161–172. https://doi.org/10.1002/jts.20194

Leen-Feldner EW, Feldner MT, Bunaciu L, Blumenthal H (2011) Associations between parental posttraumatic stress disorder and both offspring internalizing problems and parental aggression within the National Comorbidity Survey-Replication. J Anxiety Disord 25(2):169–175. https://doi.org/10.1016/j.janxdis.2010.08.017

MacDonald C, Chamberlain K, Long N, Flett R (1999) Posttraumatic stress disorder and interpersonal functioning in Vietnam War veterans: a mediational model. J Trauma Stress 12(4):701–707. https://doi.org/10.1023/A:1024729520686

Possemato K, Maisto SA, Wade M, Barrie K, McKenzie S, Lantinga LJ, Ouimette P (2015a) Ecological momentary assessment of PTSD symptoms and alcohol use in combat veterans. Psychol Addict Behav 29(4):894–905. https://doi.org/10.1037/adb0000129

Possemato K, Pratt A, Barrie K, Ouimette P (2015b) Family functioning in recent combat veterans with posttraumatic stress disorder and alcohol misuse. Traumatology 21(4):267–272. https://doi.org/10.1037/trm0000037

Rodrigues CS (2014) Dyadic examination of posttraumatic stress symptoms, relationship satisfaction, and potential mediators in military couples. PhD dissertation. University of Utah

Samper RE, Taft CT, King DW, King LA (2004) Posttraumatic stress disorder symptoms and parenting satisfaction among a national sample of male Vietnam veterans. J Trauma Stress 17(4):311–315. https://doi.org/10.1023/B:JOTS.0000038479.30903

Sullivan K, Barr N, Kintzle S, Gilreath T, Castro CA (2016) PTSD and physical health symptoms among veterans: association with child and relationship functioning. Marriage Fam Rev 52(7):689–705. https://doi.org/10.1080/01494929.2016.1157122

Taft CT, Watkins LE, Stafford J, Street AE, Monson CM (2011) Posttraumatic stress disorder and intimate relationship problems: a meta-analysis. J Consult Clin Psychol 79(1):22–33. https://doi.org/10.1037/a0022196

van Ee E, Kleber RJ, Jongmans MJ, Mooren TT, Out D (2016) Parental PTSD, adverse parenting and child attachment in a refugee sample. Attach Hum Dev 18(3):273–291. https://doi.org/10.1080/14616734.2016.1148748

Yambo T, Johnson M (2014) An integrative review of the mental health of partners of veterans with combat-related posttraumatic stress disorder. J Am Psychiatr Nurses Assoc 20(1):31–41. https://doi.org/10.1177/1078390313516998

Part III
Ameliorating Veterans' PTSD with Psychotherapies and Pharmacotherapies

Chapter 7
Psychotherapies for PTSD

7.1 Types of Psychotherapies Used to Treat PTSD

7.1.1 Cognitive Therapy and Cognitive Behaviour Therapy (CBT)

Cognitive therapy (CT) is designed to identify and modify ways of thinking that underlie major emotional concerns and distress with associated behaviours. It requires sessions with a therapist to identify these thought patterns and in so doing to recognize their problematic nature, so relieving the patient of distress. It is generally specified under the umbrella term cognitive behaviour therapy (CBT).

CBT (including CT) addresses negative thought patterns and maladaptive behaviours associated with a trauma. In general, symptom severity is reduced through cognitive restructuring and behavioural adaptation.

7.1.2 Trauma-Focused CBT

Trauma-focused CBT (TF-CBT) integrates trauma-sensitive interventions into standard CBT. This involves psychoeducation about trauma, cognitive restructuring of trauma-related beliefs and graduated exposure exercises, both imaginal and in vivo, to reduce distress associated with trauma memories. In the United Kingdom, Australia and European contexts, TF-CBT is frequently taken to include prolonged exposure (PE), cognitive processing therapy (CPT) and CT, whereas in the American context, PE, CPT and CT are stand-alone therapies.

M. Bennett, *Veteran's Stress, Trauma and Synapse Rehabilitation*,
https://doi.org/10.1007/978-981-95-9594-5_7

7.1.3 Prolonged Exposure

PE focuses on systematic confrontation with trauma-related memories and cues in a safe therapeutic context. Patients repeatedly recount their trauma (imaginal exposure) and face avoided situations (in vivo exposure) until anxiety diminishes.

7.1.4 Cognitive Processing Therapy

CPT aims to help individuals make sense of traumatic experiences, to challenge unhelpful maladaptive beliefs about the trauma (e.g. self-blame) and to manage the associated emotions that may be causing distress. CPT facilitates the development of more adaptive, balanced beliefs.

7.1.5 Eye Movement Desensitization and Reprocessing Therapy

Eye movement desensitization and reprocessing (EMDR) therapy mostly uses guided eye movements while the patient recalls traumatic memories. This procedure decreases the vividness of the memory and the emotional distress associated with it.

7.1.6 Narrative Exposure Therapy

Narrative exposure therapy (NET) involves the patient's establishment of a coherent narrative of the traumatic experiences, incorporating this into their life's narrative, focusing on addressing negative emotional experiences and rebuilding positive response patterns to them.

7.1.7 *Stress Inoculation Training*

Stress inoculation training (SIT) teaches anxiety-management skills—such as relaxation training, cognitive coping and problem-solving—to bolster resilience against trauma reminders.

7.2 Extent of Amelioration of PTSD by Psychotherapies

There has been extensive research on the relative merits of the above psychotherapies for the amelioration of PTSD. There are also a number of meta-analyses of the results of these studies. The key findings are summarized in Table 7.1.

The effectiveness of the therapies in Table 7.1 may be ranked according to the following criteria, which is used in Table 7.2:

1. Highest effect size (standardized mean difference, SMD) reported across studies, where the most negative SMD indicates the strongest evidence of symptom reduction
2. Consistently large effect sizes across multiple sources
3. Meta-analytic support (i.e. the strongest results from multiple meta-analyses)
4. Presence of direct comparisons in literature

Table 7.2 shows that the most effective psychotherapies for treating PTSD are EMDR, CPT and TF-CBT.

Table 7.1 Findings from meta-analyses on the relative merits of different psychotherapies for amelioration of PTSD

Therapy	Direct comparisons[a]/ other	SMD/WMD[b]	HG[c]
Cognitive behaviour therapy (CBT)	–	SMD: −1.09[d], −0.83[e], −1.22[f], −0.72[g]	−0.90[h]
Cognitive processing therapy (CPT)	82% better at follow-up[i]	SMD: −1.40[d], −1.30[j], −1.53[e], −1.67[g] WMD: −32.2[d]	–
Eye-movement desensitization and reprocessing (EMDR) therapy	EMDR > CBT (SMD −0.43)[k] EMDR = TF-CBT (SMD −0.17)[l]	SMD: −1.08[d], −1.10[j], −1.17[m], −1.19[g], −1.39[e], −1.51[n], −2.07[f]	−1.26[h]
Narrative exposure therapy (NET)	–	SMD: −1.25[d], 1.19[e], −0.98[g] WMD: −10.2[d]	–

(continued)

Table 7.1 (continued)

Therapy	Direct comparisons[a]/ other	SMD/WMD[b]	HG[c]
Prolonged exposure (PE) therapy	PE = CPT (SMD −0.18)[l]	SMD: −1.10[e], −1.27[g]	–
Stress inoculation training (SIT)	–	SMD: −1.07[g]	–
Trauma-focused cognitive behaviour therapy (TF-CBT)	TF-CBT = CBT (SMD −0.10)[l] TF-CBT = CBT (HG 0.14)[o] TF-CBT = EMDR[m]	SMD: −1.46[f], −1.62[m], −1.40[n]	

HG Hedges 'g', *PTSD* posttraumatic stress disorder, *SMD* standardized mean difference, *WMD* weighted mean difference

Notes:

[a]Direct comparisons in the literature only include those supported by numeric SMDs; effect size is for the difference between the therapies

[b]SMD data are Cohen's 'd' effect sizes; a small effect size is *d* d = 0.20, medium effect size is *d* d = 0.50 and large effect size is *d* d = 0.80. WMD data for PTSD symptoms are mean change from baseline in clinician-administered PTSD scale score (see Sect. 2.3.2) compared with inactive comparators. Negative WMDs and SMDs favour the intervention

[c]HG effect sizes are the same as Cohen's '*d*' for large samples (see note b)

[d]Cusack et al. (2016)

[e]Yunitri et al. (2023), Table 4

[f]Mavranezouli et al. (2020), Table 1

[g]Jericho et al. (2022), Table 4

[h]Karatzias et al. (2019)

[i]Asmundson et al. (2019)

[j]Xiang et al. (2021), Fig. 4

[k]Chen et al. (2015)

[l]Lewis et al. (2020), Table 4

[m]Bisson et al. (2013)

[n]Bisson et al. (2007)

[o]Tran and Gregor (2016)

Table 7.2 Ranking of effectiveness of psychotherapies in treating PTSD

Treatment	Best reported SMD	Range of SMDs reported	Notes
Eye-movement desensitization and reprocessing (EMDR) therapy	−2.07	−2.07 to −1.1	Highest effect size; consistently strong results
Cognitive processing therapy (CPT)	−1.67	−1.67 to −1.30	Very strong across all metrics
Trauma-focused cognitive behaviour therapy (TF-CBT)	−1.62	−1.62 to −1.46	Comparable to EMDR in some studies
Prolonged exposure (PE) therapy	−1.27	−1.27 to −1.10	Slightly lower than CPT and TF-CBT

(continued)

Table 7.2 (continued)

Treatment	Best reported SMD	Range of SMDs reported	Notes
Narrative exposure therapy (NET)	−1.25	−1.25 to −0.98	Moderate to high efficacy
Cognitive behaviour therapy (CBT)	−1.22	−1.22 to −0.72	Broad range; less effective than trauma-focused versions
Stress inoculation training (SIT)	−1.07	(Single report)	Limited data but relatively effective

Source: Table 7.1
SMD standardized mean difference

7.3 Support for Efficacy of Trauma-Focused Psychotherapies

The findings in Sect. 7.2—that the best psychotherapies are EMDR, CPT and TF-CBT, followed by PE—align with the 2018 US clinical practice guideline for the management of PTSD, which recommends manualized trauma-focused therapies as first-line interventions:

> For patients with PTSD, we recommend individual, manualized trauma-focused psychotherapies that have a primary component of exposure and/or cognitive restructuring to include Prolonged Exposure (PE), Cognitive Processing Therapy (CPT), Eye Movement Desensitization and Reprocessing (EMDR).[1]

Recent meta-analyses and systematic reviews consistently support the efficacy of trauma-focused psychotherapies for treating PTSD in adults. CPT, PE, CT and EMDR have shown large effect sizes in reducing PTSD symptoms compared to waitlist or treatment as usual.[2] TF-CBT approaches, including CPT and PE, consistently demonstrate strong evidence for effectiveness.[3] However, non-response rates also remain high, with between 60% and 72% of patients retaining their PTSD diagnoses post-treatment.[4]

[1] VA/DoD (2018), p. 431.
[2] Cusack et al. (2016), Lewis et al. (2020), and Mavranezouli et al. (2020).
[3] Schnyder (2014) and Steenkamp et al. (2015).
[4] Steenkamp et al. (2015).

7.4 Psychotherapy Amelioration of Different PTSD Symptom Clusters

The kind of treatment that a PTSD patient receives can depend on the specific PTSD symptom cluster—intrusion (B), avoidance (C), negative alterations in cognition and mood (D) and alterations in arousal and reactivity (E) (see Chap. 2)—that is most severe for that patient.

CBT has demonstrated medium to large effects on PTSD symptoms, particularly changes in arousal (E).[5] The mechanisms of CBT's effectiveness are not fully understood, with mixed findings regarding the role of physiological arousal changes during treatment.[6] CBT has also shown promise in treating PTSD arousal symptoms in adults with a history of childhood adversity. Overall, CBT remains a well-supported intervention for PTSD arousal symptoms across various populations and delivery formats.

EMDR therapy has shown effectiveness in treating PTSD intrusion symptoms (B).[7] Compared to treatments like CBT, EMDR may lead to faster recovery and greater reduction in this symptom.[8] The therapy's efficacy extends to both visual and auditory traumatic intrusions.[9] EMDR has also demonstrated potential in preventing PTSD symptoms when applied early after trauma exposure.[10]

PE therapy has shown particular efficacy in reducing avoidance (C) and cognition and mood (D) symptoms,[11] with improvements observed in both clinician-rated and self-reported PTSD measures.[12] PE also ameliorates conditions comorbid with PTSD, such as substance-use disorders, depression and suicide ideation.[13] PTSD patients cared for by Veterans Administration in the United States show recovery when treated with PE in large-scale clinical trials.[14]

TF-CBT is noted for successful treatments of children and adolescents suffering from PTSD in clinical trials.[15]

Table 7.3 provides the best evidence-based treatments for each PTSD symptom cluster, summarized as follows:

- Intrusion (B): EMDR yields marginally larger reductions in symptoms compared to CBT; CPT achieves very high recovery rates.

[5] Kuester et al. (2016) and Kar (2011).

[6] Wisco et al. (2016).

[7] Lee et al. (2002) and Chen et al. (2015).

[8] Nijdam et al. (2012) and Wilson et al. (2018).

[9] Matthijssen et al. (2017).

[10] Hagenaars and Arntz (2012).

[11] Schnurr and Lunney (2015).

[12] Schnurr and Lunney (2015) and Nishith et al. (2002).

[13] van Minnen et al. (2015).

[14] Eftekhari et al. (2013).

[15] Mavranezouli et al. (2020), Jensen et al. (2014), Table 2, and Lewey et al. (2018).

Table 7.3 Relative amelioration of the four principal PTSD symptom clusters by different psychotherapies

Psychotherapy	Intrusion (B)	Avoidance (C)	Cognition and mood (D)	Arousal and reactivity (E)
Eye-movement desensitization and reprocessing (EMDR) therapy	SMD −0.75[a]	EMDR = CBT[b] SMD −0.12[a]	–	SMD 0.39[a]
Cognitive processing therapy (CPT)	98%[c]	97%[c]	96%[c]	74%[c]
Trauma-focused cognitive behaviour therapy (TF-CBT)	SMD −0.30 (−0.21)[d]	SMD −0.36 (−0.05)[d]	SMD −0.52 (−0.02)[d]	SMD −0.61 (−0.01)[d]
Prolonged exposure (PE) therapy	SMD −0.59[e], 1.67[f]	SMD −0.77[e], 1.56[f]	SMD −0.70[e]	SMD −0.62[e], 1.34[f]
Narrative exposure therapy (NET)	–	SMD −1.13[a]	–	SMD −0.84[a]
Cognitive behaviour therapy (CBT)	SMD −0.72[a]	SMD −0.68[a]	–	SMD −1.30[a]

SMD standardized mean difference

Notes:

[a]Gillies et al. (2013)

[b]Chen et al. (2015), Figs. 4, 5, and 6

[c]Values are % decreases of symptom cluster in those that showed recovery: from Miles et al. (2023), Table 3

[d]SMD values are for intake to post-treatment (from post-treatment to 3-month follow-up); effect sizes following TF-CBT likely to be a mixture of PE, CPT, EMDR and more general trauma-focused cognitive therapy; note the decline in effectiveness between post-treatment and 3-month follow-up: from Phelps et al. (2018), Table 2

[e]Effect sizes for pre–post-treatment of female veterans; note greater effect on C and D clusters: from Schnurr and Lunney (2015)

[f]Horesh et al. (2017)

- Avoidance (C): PE and NET demonstrate medium to large effects; CPT again shows marked improvements.
- Negative alterations in cognition and mood (D): PE demonstrates large effects; CPT maintains high efficacy.
- Alterations in arousal and reactivity (E): CBT and NET show large effects; PE demonstrates sustained arousal reductions.

References

Asmundson GJG, Thorisdottir AS, Roden-Foreman JW, Baird SO, Witcraft SM, Stein AT, Smits JAJ, Powers MB (2019) A meta-analytic review of cognitive processing therapy for adults with posttraumatic stress disorder. Cogn Behav Ther (1):1–14. https://doi.org/10.1080/16506073.2018.1522371

Bisson JI, Ehlers A, Matthews R, Pilling S, Richards D, Turner S (2007) Psychological treatments for chronic post-traumatic stress disorder: systematic review and meta-analysis. Br J Psychiatry 190:97–104. https://doi.org/10.1192/bjp.bp.106.021402

Bisson JI, Roberts NP, Andrew M, Cooper R, Lewis C (2013) Psychological therapies for chronic post-traumatic stress disorder (PTSD) in adults. Cochrane Database Syst Rev 2013(12):CD003388. https://doi.org/10.1002/14651858.cd003388.pub4

Chen L, Zhang G, Hu M, Liang X (2015) Eye movement desensitization and reprocessing versus cognitive-behavioral therapy for adult posttraumatic stress disorder: systematic review and meta-analysis. J Nerv Ment Dis 203(6):443–451. https://doi.org/10.1097/nmd.0000000000000306

Cusack K, Jonas DE, Forneris CA, Wines C, Sonis J, Middleton JC, Feltner C, Brownley KA, Olmsted KR, Greenblatt A, Weil A, Gaynes BN (2016) Psychological treatments for adults with posttraumatic stress disorder: a systematic review and meta-analysis. Clin Psychol Rev 43:128–141. https://doi.org/10.1016/j.cpr.2015.10.003

Eftekhari A, Ruzek JI, Crowley JJ, Rosen CS, Greenbaum MA, Karlin BE (2013) Effectiveness of national implementation of prolonged exposure therapy in Veterans Affairs care. JAMA Psychiatry 70(9):949–955. https://doi.org/10.1001/jamapsychiatry.2013.36

Gillies D, Taylor F, Gray C, O'Brien L, D'Abrew N (2013) Psychological therapies for the treatment of post-traumatic stress disorder in children and adolescents (review). Evid Based Child Health 8(3):1004–1116. https://doi.org/10.1002/ebch.1916

Hagenaars MA, Arntz A (2012) Reduced intrusion development after post-trauma imagery rescripting; an experimental study. J Behav Ther Exp Psychiatry 43(2):808–814. https://doi.org/10.1016/j.jbtep.2011.09.005

Horesh D, Qian M, Freedman S, Shalev A (2017) Differential effect of exposure-based therapy and cognitive therapy on post-traumatic stress disorder symptom clusters: a randomized controlled trial. Psychol Psychother 90(2):235–243. https://doi.org/10.1111/papt.12103

Lee HA, Gabriel R, Bolton JP, Bale AJ, Jackson M (2002) Health status and clinical diagnoses of 3000 UK Gulf War veterans. J R Soc Med 95(10):491–497. https://doi.org/10.1177/014107680209501004

Jensen TK, Holt T, Ormhaug SM, Egeland K, Granly L, Hoaas LC, Hukkelberg SS, Indregard T, Stormyren SD, Wentzel-Larsen T (2014) A randomized effectiveness study comparing trauma-focused cognitive behavioral therapy with therapy as usual for youth. J Clin Child Adolesc Psychol 43(3):356–369. https://doi.org/10.1080/15374416.2013.822307

Jericho B, Luo A, Berle D (2022) Trauma-focused psychotherapies for post-traumatic stress disorder: a systematic review and network meta-analysis. Acta Psychiatr Scand 145(2):132–155. https://doi.org/10.1111/acps.13366

Kar N (2011) Cognitive behavioral therapy for the treatment of post-traumatic stress disorder: a review. Neuropsychiatr Dis Treat 7:167–181. https://doi.org/10.2147/ndt.s10389

Karatzias T, Murphy P, Cloitre M, Bisson J, Roberts N, Shevlin M, Hyland P, Maercker A, Ben-Ezra M, Coventry P, Mason-Roberts S, Bradley A, Hutton P (2019) Psychological interventions for ICD-11 complex PTSD symptoms: systematic review and meta-analysis. Psychol Med 49(11):1761–1775. https://doi.org/10.1017/s0033291719000436

Kuester A, Niemeyer H, Knaevelsrud C (2016) Internet-based interventions for posttraumatic stress: a meta-analysis of randomized controlled trials. Clin Psychol Rev 43:1–16. https://doi.org/10.1016/j.cpr.2015.11.004

Lewey JH, Smith CL, Burcham B, Saunders NL, Elfallal D, O'Toole SK (2018) Comparing the effectiveness of EMDR and TF-CBT for children and adolescents: a meta-analysis. J Child Adolesc Trauma 11(4):457–472. https://doi.org/10.1007/s40653-018-0212-1

Lewis C, Roberts NP, Andrew M, Starling E, Bisson JI (2020) Psychological therapies for post-traumatic stress disorder in adults: systematic review and meta-analysis. Eur J Psychotraumatol 11(1):1729633. https://doi.org/10.1080/20008198.2020.1729633

Matthijssen SJMA, Verhoeven LCM, van den Hout MA, Heitland I (2017) Auditory and visual memories in PTSD patients targeted with eye movements and counting: the effect of modality-specific loading of working memory. Front Psychol 8:1937. https://doi.org/10.3389/fpsyg.2017.01937

Mavranezouli I, Megnin-Viggars O, Daly C, Dias S, Stockton S, Meiser-Stedman R, Trickey D, Pilling S (2020) Research review: psychological and psychosocial treatments for children and

young people with post-traumatic stress disorder: a network meta-analysis. J Child Psychol Psychiatry 61(1):18–29. https://doi.org/10.1111/jcpp.13094
Miles SR, Hale WJ, Mintz J, Wachen JS, Litz BT, Dondanville KA, Yarvis JS, Hembree EA, Young-McCaughan S, Peterson AL, Resick PA (2023) Hyperarousal symptoms linger after successful PTSD treatment in active duty military. Psychol Trauma 15(8):1398–1405. https://doi.org/10.1037/tra0001292
Nijdam MJ, Gersons BP, Reitsma JB, de Jongh A, Olff M (2012) Brief eclectic psychotherapy v. eye movement desensitisation and reprocessing therapy for post-traumatic stress disorder: randomised controlled trial. Br J Psychiatry 200(3):224–231. https://doi.org/10.1192/bjp.bp.111.099234
Nishith P, Resick PA, Griffin MG (2002) Pattern of change in prolonged exposure and cognitive-processing therapy for female rape victims with posttraumatic stress disorder. J Consult Clin Psychol 70(4):880–886. https://doi.org/10.1037/0022-006x.70.4.880
Phelps AJ, Steele Z, Cowlishaw S, Metcalf O, Alkemade N, Elliott P, O'Donnell M, Redston S, Kerr K, Howard A, Nursey J, Cooper J, Armstrong R, Fitzgerald L, Forbes D (2018) Treatment outcomes for military veterans with posttraumatic stress disorder: response trajectories by symptom cluster. J Trauma Stress 31(3):401–409. https://doi.org/10.1002/jts.22299
Schnurr PP, Lunney CA (2015) Differential effects of prolonged exposure on posttraumatic stress disorder symptoms in female veterans. J Consult Clin Psychol 83(6):1154–1160. https://doi.org/10.1037/ccp0000031
Schnyder U (2014) Treating intrusions, promoting resilience: an overview of therapies for trauma-related psychological disorders. Eur J Psychotraumatol 5:26520. https://doi.org/10.3402/ejpt.v5.26520
Steenkamp MM, Litz BT, Hoge CW, Marmar CR (2015) Psychotherapy for military-related PTSD: a review of randomized clinical trials. JAMA 314(5):489–500. https://doi.org/10.1001/jama.2015.8370
Tran US, Gregor B (2016) The relative efficacy of bona fide psychotherapies for post-traumatic stress disorder: a meta-analytical evaluation of randomized controlled trials. BMC Psychiatry 16:266. https://doi.org/10.1186/s12888-016-0979-2
VA/DOD Clinical practice guideline for the management of posttraumatic stress disorder and acute stress disorder: clinician summary (2018) Focus (Am Psychiatr Publ) 16(4):430–448. https://doi.org/10.1176/appi.focus.16408
van Minnen A, Zoellner LA, Harned MS, Mills K (2015) Changes in comorbid conditions after prolonged exposure for PTSD: a literature review. Curr Psychiatry Rep 17(3):549. https://doi.org/10.1007/s11920-015-0549-1
Wilson G, Farrell D, Barron I, Hutchins J, Whybrow D, Kiernan MD (2018) The use of eye-movement desensitization reprocessing (EMDR) therapy in treating post-traumatic stress disorder—a systematic narrative review. Front Psychol 9:923. https://doi.org/10.3389/fpsyg.2018.00923
Wisco BE, Marx BP, Miller MW, Wolf EJ, Mota NP, Krystal JH, Southwick SM, Pietrzak RH (2016) Probable posttraumatic stress disorder in the US veteran population according to DSM-5: results from the National Health and Resilience in Veterans Study. J Clin Psychiatry 77(11):1503–1510. https://doi.org/10.4088/jcp.15m10188
Xiang Y, Cipriani A, Teng T, Del Giovane C, Zhang Y, Weisz JR, Li X, Cuijpers P, Liu X, Barth J, Jiang Y, Cohen D, Fan L, Gillies D, Du K, Ravindran AV, Zhou X, Xie P (2021) Comparative efficacy and acceptability of psychotherapies for post-traumatic stress disorder in children and adolescents: a systematic review and network meta-analysis. Evid Based Ment Health 24(4):153–160. https://doi.org/10.1136/ebmental-2021-300346
Yunitri N, Chu H, Kang XL, Wiratama BS, Lee TY, Chang LF, Liu D, Kustanti CY, Chiang KJ, Chen R, Tseng P, Chou KR (2023) Comparative effectiveness of psychotherapies in adults with posttraumatic stress disorder: a network meta-analysis of randomised controlled trials. Psychol Med 53(13):6376–6388. https://doi.org/10.1017/s0033291722003737

Chapter 8
Pharmacotherapies for PTSD

8.1 Types of Pharmacotherapies Used to Manage PTSD

Pharmacological therapies play an important role in managing PTSD, particularly when psychotherapy is not feasible or is insufficient. Selective serotonin reuptake inhibitors (SSRIs) and serotonin–norepinephrine reuptake inhibitors (SNRIs) are the most widely used pharmacological treatments, offering moderate improvements across all PTSD symptom clusters (see Sect. 2.1). Ketamine is highly effective for specific PTSD symptom clusters, like hyperarousal, re-experiencing and negative moods, and 3,4-methylenedioxymethamphetamine (MDMA) for the symptom clusters of re-experiencing, avoidance and negative moods. The effectiveness of each pharmacotherapy treatment depends on the patient's specific PTSD symptom profile.

8.1.1 Selective Serotonin Reuptake Inhibitors

SSRIs increase serotonin levels in the brain, which helps regulate mood, anxiety and emotional responses.[1] The SSRIs sertraline and paroxetine effectively reduce the core symptoms of PTSD in some patients, although less so for avoidance (Table 8.2). Indeed about 40% of patients do not respond to SSRIs at all.[2]

---

[1] Martin et al. (2021), Cukor et al. (2010).

[2] Bisson and Andrew (2007), Bisson et al. (2015).

M. Bennett, *Veteran's Stress, Trauma and Synapse Rehabilitation*,
https://doi.org/10.1007/978-981-95-9594-5_8

8.1.2 Serotonin–Norepinephrine Reuptake Inhibitors

SNRIs provide an alternative by increasing both serotonin and norepinephrine levels (Table 8.1).[3] Venlafaxine has been shown to be effective in some cases when SSRIs have failed, improving symptoms related to hyperarousal and intrusive thoughts (Table 8.2).[4] However, as with SSRIs, avoidance behaviours remain relatively resistant to SNRIs. The therapeutic effect of SNRIs also takes several weeks to manifest, making them less suitable for acute symptom management but valuable for long-term maintenance.

8.1.3 Prazosin

Prazosin, an alpha-1 adrenergic antagonist, is used to treat PTSD-related nightmares and sleep disturbances. It has significant efficacy in reducing the frequency and severity of nightmares (Table 8.1),[5] although its overall impact on PTSD symptoms is more limited (Table 8.2).[6]

8.1.4 Benzodiazepines

Benzodiazepines, such as clonazepam and lorazepam, are sometimes used to manage hyperarousal, anxiety and insomnia in PTSD patients (Tables 8.1 and 8.2). However, their long-term use is generally discouraged due to the risk of dependence[7] and they are not, or should not be considered, a frontline treatment for PTSD.[8]

8.1.5 Ketamine

Ketamine modulates the action of glutamate, a neurotransmitter involved in memory and learning, as a consequence of its antagonism of the glutamate NMDA receptor. Intravenous ketamine infusions have been shown to produce rapid reductions in PTSD symptoms, particularly in hyperarousal, negative mood and

---

[3] Davis et al. (2001, 2006).

[4] Bajor et al. (2022).

[5] Schrader and Ross (2021), Raskind et al. (2007).

[6] Bisson et al. (2015).

[7] Martényi (2005), Davis et al. (2006), Shalev et al. (1996).

[8] Bisson and Andrew (2007), Bisson et al. (2015).

Table 8.1 Relative amelioration of PTSD by different pharmacological treatments

Pharmacological therapy	Description of effect	Duration of effect after session	Extent of PTSD amelioration
SSRIs (e.g. sertraline and paroxetine)	Increase serotonin levels to regulate mood and anxiety	Several months to years, depending on adherence, with full adherence improving less than 50% of patients[a]	Good
SNRIs (e.g. venlafaxine)	Affect both serotonin and norepinephrine to improve mood, re-experiencing and arousal	Several months to years, with continued use[b]	Moderate
Prazosin	Alpha-1 adrenergic antagonist; primarily used to reduce nightmares and improve sleep quality	Effects can last throughout nightly administration but cease after discontinuation[c]	Moderate (for sleep-related symptoms)
Benzodiazepines (e.g. lorazepam and diazepam)	Anxiolytics to reduce anxiety	Short-term; effects last a few hours[d]	Fair
Ketamine	NMDA receptor antagonist; shown to have rapid effects on reducing hyperarousal, re-experiencing and mood	1–2 weeks after a single dose, longer with repeated administration[e]	Moderate to good
Antipsychotics (e.g. quetiapine)	Treats severe mood disturbances and psychotic symptoms	Several weeks to months, depending on the treatment[f]	Fair (as adjunctive therapy)
MDMA	Entactogen with stimulant and minor psychedelic properties	Long-term effects noted, particularly when combined with psychotherapy[g]	Moderate to good

MDMA 3,4-methylenedioxymethamphetamine, *PTSD* posttraumatic stress disorder, *SNRIs* serotonin-norepinephrine reuptake inhibitors, *SSRIs* selective serotonin reuptake inhibitors
[a]Cukor et al. (2010)
[b]Davidson (2006)
[c]Raskind et al. (2007)
[d]Martényi (2002)
[e]Feder et al. (2014), Krystal et al. (2017)
[f]Adetunji et al. (2005)
[g]Mithoefer et al. (2011)

re-experiencing, often within hours to days after administration (Table 8.2).[9] Ketamine's rapid onset of action represents a significant advantage over traditional agents, which may take weeks to show effect (Table 8.1). It has to be managed carefully with individual patients because of its tendency to induce dissociation and hallucinations. Ketamine is highly effective in cases of treatment-resistant PTSD,

[9] Feder et al. (2014), Davis et al. (2006).

Table 8.2 Relative amelioration of the four principal PTSD symptom clusters by different pharmacological treatments

Pharmacological therapy	Intrusion (B)	Avoidance (C)	NACM (D)	AAR (E)
SSRIs (e.g. sertraline and paroxetine)[a,b]	Fair to moderate	Fair	Moderate	Fair to moderate
SNRIs (e.g. venlafaxine)[b]	Fair to moderate	Fair	Moderate	Moderate
Prazosin[c,d]	Good	Poor	Poor	Moderate
Benzodiazepines (e.g. lorazepam and diazepam)[e]	Poor to fair	Poor	Poor	Fair
Ketamine[f,g]	Moderate to good	Poor	Moderate to good	Moderate to good
Antipsychotics (e.g. quetiapine)[h]	Poor	Poor	Moderate	Moderate
MDMA[i]	Moderate to good	Moderate to good	Moderate to good	Moderate

AAR alterations in arousal and reactivity, *MDMA* 3,4-methylenedioxymethamphetamine, *NACM* negative alterations in cognition and mood, *PTSD* posttraumatic stress disorder
[a]Brady et al. (2000)
[b]Davidson (2006)
[c]Raskind et al. (2007)
[d]Taylor et al. (2008)
[e]Martényi (2002)
[f]Feder et al. (2014)
[g]Krystal et al. (2017)
[h]Sharpless and Barber (2011)
[i]Mithoefer et al. (2011)

showing substantial improvements in mood and anxiety.[10] Ketamine is coming to play a critical role in managing PTSD.[11]

8.1.6 *Antipsychotics and Mood Stabilizers*

Mood stabilizers and antipsychotics are only prescribed as adjuncts to the principal pharmacological therapy agitation. For example, PTSD patients with very severe mood instabilities may benefit from an adjunct mood stabilizer, such as lamotrigine. Likewise, patients with dissociative PTSD may benefit from adjunct antipsychotic treatment with quetiapine (Tables 8.1 and 8.2). Adjunct mood stabilizers and antipsychotics are useful in reducing hyperarousal symptoms and negative mood and cognition in patients with comorbid disorders (Table 8.2).[12]

[10] Feder et al. (2014).
[11] Mello et al. (2021).
[12] Sharpless and Barber (2011).

8.1.7 3,4-Methylenedioxymethamphetamine

MDMA has powerful effects in ameliorating the PTSD symptoms of re-experiencing, avoidance and arousal (Table 8.2). It is often used in conjunction with cognitive behavioural therapy (see Chap. 7), especially among individuals with chronic treatment-resistant PTSD.[13]

8.2 Extent of Amelioration of PTSD by Pharmacotherapies

Table 8.1 presents the results of studies into the extent that specific pharmacological therapies have had on ameliorating PTSD among the general population. Only one study, on prazosin, was veteran specific. The studies suggest that SSRIs, ketamine and MDMA are the most effective for PTSD, when used within the appropriate clinical parameters.

8.3 Pharmacotherapy Amelioration of Different PTSD Symptom Clusters

Table 8.2 presents the results of studies into the effects of each pharmacotherapy on each of the four PTSD symptom clusters: intrusion (B), avoidance (C), negative alterations in cognition and mood (D) and alterations in arousal and reactivity (E). As for the studies in Table 8.1, only one of the prazosin studies was specific to veterans. The results show that for intrusion symptoms, most pharmacotherapies have some effect, with only benzodiazepines and antipsychotics at the poor end of the scale, while only MDMA is effective for ameliorating avoidance symptoms. For cognition and mood symptoms, most of the pharmacotherapies have some effect, the best being ketamine and MDMA, but prazosin and benzodiazepines are poor treatments of these symptoms. For arousal and reactivity symptoms, the best treatment is ketamine, with all others showing fair to moderate amelioration. The studies highlight the need for personalized psychopharmacological treatments for PTSD patients.

[13] Mithoefer et al. (2011).

References

Adetunji B, Mathews M, Williams A, Budur K, Mathews M, Mahmud J, Osinowo T (2005) Use of antipsychotics in the treatment of post-traumatic stress disorder. Psychiatry 2(4):43–47

Bajor LA, Balsara C, Osser DN (2022) An evidence-based approach to psychopharmacology for posttraumatic stress disorder (PTSD)—2022 update. Psychiatry Res 317:114840. https://doi.org/10.1016/j.psychres.2022.114840

Bisson J, Andrew M (2007) Psychological treatment of post-traumatic stress disorder (PTSD). Cochrane Database Syst Rev 18(3):CD003388. Update in: Cochrane Database Syst Rev 2013 Dec 13;(12):CD003388. https://doi.org/10.1002/14651858.cd003388.pub3

Bisson JI, Cosgrove S, Lewis C, Roberts NP (2015) Post-traumatic stress disorder. BMJ 351:h6161. https://doi.org/10.1136/bmj.h6161

Brady K, Pearlstein T, Asnis GM, Baker D, Rothbaum B, Sikes CR, Farfel GM (2000) Efficacy and safety of sertraline treatment of posttraumatic stress disorder: a randomized controlled trial. JAMA 283(14):1837–1844. https://doi.org/10.1001/jama.283.14.1837

Cukor J, Olden M, Lee F, Difede J (2010) Evidence-based treatments for PTSD, new directions, and special challenges. Ann N Y Acad Sci 1208:82–89. https://doi.org/10.1111/j.1749-6632.2010.05793.x

Davidson JR (2006) Pharmacologic treatment of acute and chronic stress following trauma: 2006. J Clin Psychiatry 67(Suppl 2):34–39

Davis LL, English BA, Ambrose SM, Petty F (2001) Pharmacotherapy for post-traumatic stress disorder: a comprehensive review. Expert Opin Pharmacother 2(10):1583–1595. https://doi.org/10.1517/14656566.2.10.1583

Davis LL, Frazier EC, Williford RB, Newell JM (2006) Long-term pharmacotherapy for post-traumatic stress disorder. CNS Drugs 20(6):465–476. https://doi.org/10.2165/00023210-200620060-00003

Feder A, Parides MK, Murrough JW, Perez AM, Morgan JE, Saxena S, Kirkwood K, Aan Het Rot M, Lapidus KA, Wan LB, Iosifescu D, Charney DS (2014) Efficacy of intravenous ketamine for treatment of chronic posttraumatic stress disorder: a randomized clinical trial. JAMA Psychiatry 71(6):681–688. https://doi.org/10.1001/jamapsychiatry.2014.62

Krystal JH, Abdallah CG, Averill LA, Kelmendi B, Harpaz-Rotem I, Sanacora G, Southwick SM, Duman RS (2017) Synaptic loss and the pathophysiology of PTSD: implications for ketamine as a prototype novel therapeutic. Curr Psychiatry Rep 19(10):74. https://doi.org/10.1007/s11920-017-0829-z

Martényi F (2005) A poszttraumás-stressz zavar gyógyszeres kezelésének három paradigmája [Three paradigms in the treatment of posttraumatic stress disorder]. Neuropsychopharmacol Hung 7(1):11–21

Martin A, Naunton M, Kosari S, Peterson G, Thomas J, Christenson JK (2021) Treatment guidelines for PTSD: a systematic review. J Clin Med 10(18):4175. https://doi.org/10.3390/jcm10184175

Mello RP, Echegaray MVF, Jesus-Nunes AP, Leal GC, Magnavita GM, Vieira F, Caliman-Fontes AT, Telles M, Guerreiro-Costa LNF, Souza-Marques B, Bandeira ID, Santos-Lima C, Marback RF, Correia-Melo FS, Lacerda ALT, Quarantini LC (2021) Trait dissociation as a predictor of induced dissociation by ketamine or esketamine in treatment-resistant depression: secondary analysis from a randomized controlled trial. J Psychiatr Res 138:576–583. https://doi.org/10.1016/j.jpsychires.2021.05.014

Mithoefer MC, Wagner MT, Mithoefer AT, Jerome L, Doblin R (2011) The safety and efficacy of {+/−}3,4−methylenedioxymethamphetamine-assisted psychotherapy in subjects with chronic, treatment-resistant posttraumatic stress disorder: the first randomized controlled pilot study. J Psychopharmacol 25(4):439–452. https://doi.org/10.1177/0269881110378371

Raskind MA, Peskind ER, Hoff DJ, Hart KL, Holmes HA, Warren D, Shofer J, O'Connell J, Taylor F, Gross C, Rohde K, McFall ME (2007) A parallel group placebo controlled study of prazosin for trauma nightmares and sleep disturbance in combat veterans with post-traumatic stress disorder. Biol Psychiatry 61(8):928–934. https://doi.org/10.1016/j.biopsych.2006.06.032

Schrader C, Ross A (2021) A review of PTSD and current treatment strategies. Mo Med 118(6):546–551

Shalev AY, Bonne O, Eth S (1996) Treatment of posttraumatic stress disorder: a review. Psychosom Med 58(2):165–182. https://doi.org/10.1097/00006842-199603000-00012

Sharpless BA, Barber JP (2011) A clinician's guide to PTSD treatments for returning veterans. Prof Psychol Res Pr 42(1):8–15. https://doi.org/10.1037/a0022351

Taylor FB, Martin P, Thompson C, Williams J, Mellman TA, Gross C, Peskind ER, Raskind MA (2008) Prazosin effects on objective sleep measures and clinical symptoms in civilian trauma posttraumatic stress disorder: a placebo-controlled study. Biol Psychiatry 63(6):629–632. https://doi.org/10.1016/j.biopsych.2007.07.001

Chapter 9
Impact of Social Support on Veterans' PTSD Symptoms

9.1 The Effect of Social Support in Alleviating PTSD

Social support includes emotional/informational support (e.g. showing care and providing guidance), positive social interaction, affectionate support and tangible support. Social support, particularly from a partner, is related to the success of PTSD treatment outcomes.[1] For example, increased social support prior to cognitive processing therapy is especially efficacious in lowering clinician-administered PTSD scores (CAPS) in response to the treatment.[2] Patients with social support on a given day have lowered PTSD symptoms.[3] Such support is most efficacious when perceived to be provided by partners or veteran peers rather than non-veteran friends and relatives.[4]

9.2 Correlations Between Social Support and PTSD Treatment Outcomes

Quantitative estimates of increased social support from family, friends and colleagues during treatment for PTSD are correlated with enhanced reduction in PTSD symptoms over PTSD treatment alone (Fig. 9.1).[5] There is an overall effect size of increased social support on PTSD relief of $r = -0.27$.[6] Increased social interactions

[1] Thrasher et al. (2010) and Guay et al. (2011).

[2] Campbell et al. (2020).

[3] Dworkin et al. (2018).

[4] Laffaye et al. (2008).

[5] Barrett and Mizes (1988), Price et al. (2018) and Stanley et al. (2019).

[6] Zalta et al. (2021).

M. Bennett, *Veteran's Stress, Trauma and Synapse Rehabilitation*,
https://doi.org/10.1007/978-981-95-9594-5_9

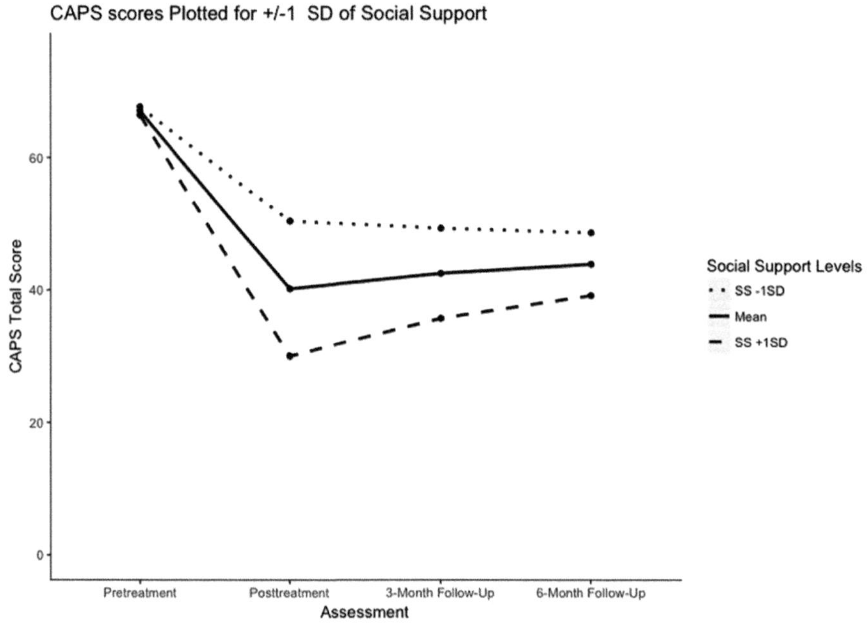

Fig. 9.1 Estimated scores on the clinician-administered PTSD scale (CAPS) for those with social support ±1 standard deviation (SD) across all assessment points. The treatment involved eight weekly 1.5-h individual sessions of prolonged exposure therapy. (Reproduced from Price et al. (2018), fig. 1, with permission from Guilford Publications)

decrease pre-treatment PTSD symptom intensity, and emotional/informational support accelerates the response to the PTSD treatment (Fig. 9.2).[7] Affectionate support and tangible support were without effect on the PTSD treatment outcome.[8] For veterans, those who received social support during deployment or post-deployment had significantly lower levels of PTSD than those who received such support pre-deployment only.[9] This was especially the case if emotional support came from family and friends ($\beta = 0.7$) but also if such support came from the community and workplace ($\beta = 0.57$). In the case of female veterans, social support from co-workers and friends is even more important than that of the family in assisting PTSD treatment outcomes ($r = -0.33$ compared with $r = -0.28$).[10]

Therapy for PTSD is associated with a perceived increase of social support by the patient.[11] The histogram below shows the extent to which PTSD patients who received prolonged exposure (PE) therapy perceived an increased level of social

[7] Price et al. (2013).

[8] Price et al. (2013).

[9] Han et al. (2014).

[10] Klarić et al. (2008).

[11] Bourassa et al. (2020).

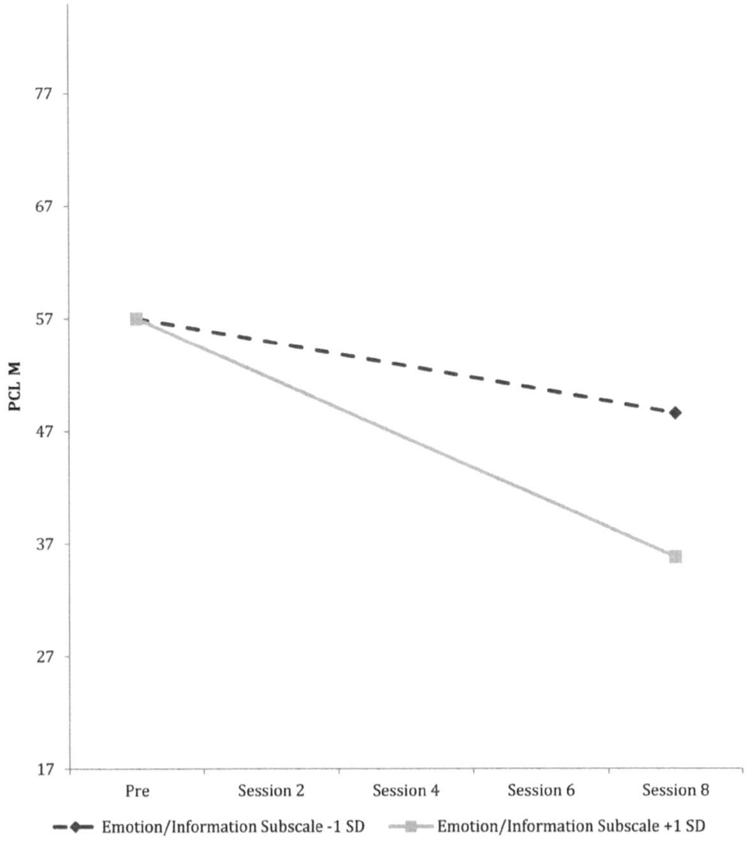

Fig. 9.2 Outcome trajectories on the PTSD checklist—military version (PCL M) for individuals scoring ±1 standard deviation (SD) on the emotional/informational subscale of the MOS Social Support Survey. The lowest possible score on the PCL M is 17. (Reproduced from Price et al. (2013), fig. 1, with permission from the American Psychological Association)

support according to the Brief Assessment of Social Support (BASIS) subscale (which includes getting along with those in and outside of the family, getting along in social situations, perceived closeness to someone and feeling they could rely on another in a time of need). Scores on the BASIS measure were recoded so that higher scores corresponded to greater perceived social support (Fig. 9.3). Indeed, social support and PTSD reciprocally predict each other ($r = 0.24$).[12]

[12] Wang et al. (2021).

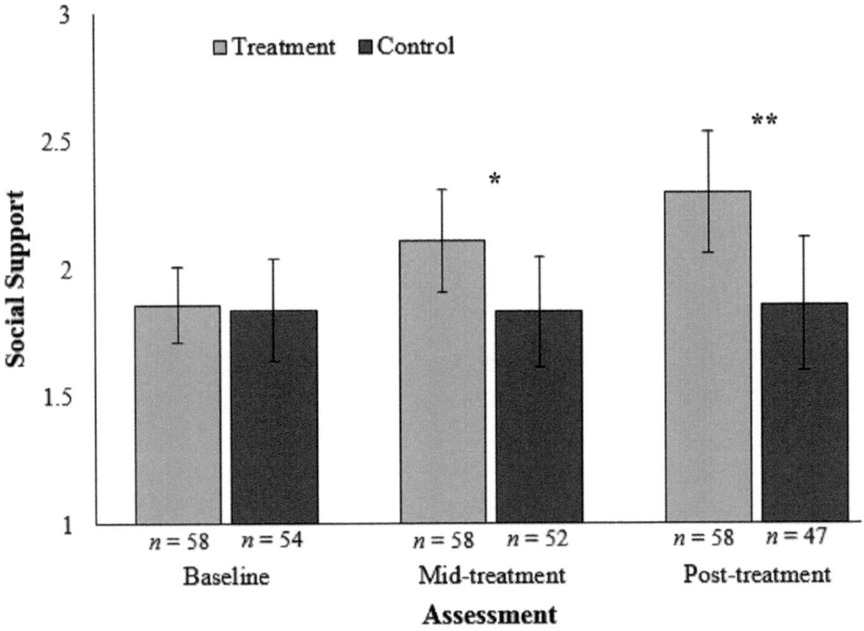

Fig. 9.3 Prolonged exposure (PE) treatment effect on social support at each occasion for control and treatment (assisted PE) participants. The full Brief Assessment of Social Support (BASIS) subscale ranges from 0 to 4 (*$p < 0.05$, **$p < 0.01$). (Reproduced from Bourassa et al. (2020), fig. 1, with permission from Elsevier)

9.3 Impact of Social Support on PTSD Symptom Clusters

Less social support has been found to lead to increases in three out of the four major PTSD symptom clusters, namely, intrusions, avoidance and hypervigilance (see Sect. 2.1).[13]

References

Barrett TW, Mizes JS (1988) Combat level and social support in the development of post-traumatic stress disorder in Vietnam veterans. Behav Modif 12(1):100–115. https://doi.org/10.1177/01454455880121005

Bourassa KJ, Smolenski DJ, Edwards-Stewart A, Campbell SB, Reger GM, Norr AM (2020) The impact of prolonged exposure therapy on social support and PTSD symptoms. J Affect Disord 260:410–417. https://doi.org/10.1016/j.jad.2019.09.036

Campbell SB, Erbes C, Grubbs K, Fortney J (2020) Social support moderates the association between posttraumatic stress disorder treatment duration and treatment outcomes in

[13] Shallcross et al. (2016).

telemedicine-based treatment among rural veterans. J Trauma Stress 33(4):391–400. https://doi.org/10.1002/jts.22542

Dworkin ER, Ullman SE, Stappenbeck C, Brill CD, Kaysen D (2018) Proximal relationships between social support and PTSD symptom severity: a daily diary study of sexual assault survivors. Depress Anxiety 35(1):43–49. https://doi.org/10.1002/da.22679

Guay S, Beaulieu-Prévost D, Beaudoin C, St-Jean-Trudel E, Nachar N, Marchand A, O'Connor KP (2011) How do social interactions with a significant other affect PTSD symptoms? An empirical investigation with a clinical sample. J Aggress Maltreat Trauma 20(3):280–303. https://doi.org/10.1080/10926771.2011.562478

Han SC, Castro F, Lee LO, Charney ME, Marx BP, Brailey K, Proctor SP, Vasterling JJ (2014) Military unit support, postdeployment social support, and PTSD symptoms among active duty and National Guard soldiers deployed to Iraq. J Anxiety Disord 28(5):446–453. https://doi.org/10.1016/j.janxdis.2014.04.004

Klarić M, Francisković T, Klarić B, Kresić M, Grković J, Lisica ID, Stevanović A (2008) Social support and PTSD symptoms in war-traumatized women in Bosnia and Herzegovina. Psychiatr Danub 20(4):466–473

Laffaye C, Cavella S, Drescher K, Rosen C (2008) Relationships among PTSD symptoms, social support, and support source in veterans with chronic PTSD. J Trauma Stress 21(4):394–401. https://doi.org/10.1002/jts.20348

Price M, Gros DF, Strachan M, Ruggiero KJ, Acierno R (2013) The role of social support in exposure therapy for operation Iraqi freedom/operation enduring freedom veterans: a preliminary investigation. Psychol Trauma 5(1):93–100. https://doi.org/10.1037/a0026244

Price M, Lancaster CL, Gros DF, Legrand AC, van Stolk-Cooke K, Acierno R (2018) An examination of social support and PTSD treatment response during prolonged exposure. Psychiatry 81(3):258–270. https://doi.org/10.1080/00332747.2017.1402569

Shallcross SL, Arbisi PA, Polusny MA, Kramer MD, Erbes CR (2016) Social causation versus social erosion: comparisons of causal models for relations between support and PTSD symptoms. J Trauma Stress 29(2):167–175. https://doi.org/10.1002/jts.22086

Stanley IH, Hom MA, Chu C, Dougherty SP, Gallyer AJ, Spencer-Thomas S, Shelef L, Fruchter E, Comtois KA, Gutierrez PM, Sachs-Ericsson NJ, Joiner TE (2019) Perceptions of belongingness and social support attenuate PTSD symptom severity among firefighters: a multistudy investigation. Psychol Serv 16(4):543–555. https://doi.org/10.1037/ser0000240

Thrasher S, Power M, Morant N, Marks I, Dalgleish T (2010) Social support moderates outcome in a randomized controlled trial of exposure therapy and (or) cognitive restructuring for chronic posttraumatic stress disorder. Can J Psychiatr 55(3):187–190. https://doi.org/10.1177/070674371005500311

Wang Y, Chung MC, Wang N, Yu X, Kenardy J (2021) Social support and posttraumatic stress disorder: a meta-analysis of longitudinal studies. Clin Psychol Rev 85:101998. https://doi.org/10.1016/j.cpr.2021.101998

Zalta AK, Tirone V, Orlowska D, Blais RK, Lofgreen A, Klassen B, Held P, Stevens NR, Adkins E, Dent AL (2021) Examining moderators of the relationship between social support and self-reported PTSD symptoms: a meta-analysis. Psychol Bull 147(1):33–54. https://doi.org/10.1037/bul0000316

Part IV
Veterans' PTSD, Grey Matter Changes and Synapse Loss

Chapter 10
Cortical Grey Matter Loss Accompanying PTSD and Its Consequences

10.1 Relationship Between Trauma, PTSD and Grey Matter Loss

Numerous region of interest (ROI) and voxel-based morphometry (VBM) studies have investigated grey matter volume (GMV) differences in patients with posttraumatic stress disorder (PTSD), trauma-exposed controls without PTSD and healthy controls. Table 10.1 compiles the principal findings.

Figure 10.1 shows grey matter loss in those with PTSD and those exposed to trauma without PTSD. The key regions with overlapping grey matter loss are the hippocampus, the anterior cingulate cortex (ACC) and the prefrontal cortex (PFC). These regions consistently show grey matter loss in both PTSD and trauma-exposed groups, suggesting that these areas are vulnerable to trauma exposure, with or without PTSD.

There are regions where grey matter loss is found predominantly in PTSD patients and is absent or less pronounced in trauma-exposed individuals without PTSD. Such regions are the subgenual ACC (sgACC),[1] amygdala[2] and middle frontal gyrus (MFG).[3] These regions indicate that while some brain areas may be affected in trauma-exposed individuals, others, such as the sgACC and amygdala, appear to be more specific to PTSD.

[1] Herringa et al. (2012).

[2] Morey et al. (2012).

[3] Herringa et al. (2012).

M. Bennett, *Veteran's Stress, Trauma and Synapse Rehabilitation*,
https://doi.org/10.1007/978-981-95-9594-5_10

Table 10.1 Findings of studies into grey matter loss in the adult human cortex accompanying PTSD

Site of grey matter loss	Combat related	Comparative results	Correlation with CAPS/ CES	Meta-analysis/ review	Authors/ sources
H; H; lH	No	T < C; P < C; P < T	H loss covaries with CAPS	Yes	Woon et al. (2010)
rH, rA	Yes	P < C		No	Akiki et al. (2017)
H; H, lA; H, ACC,	Yes	T < C; P < C; P < T		Yes	Karl et al. (2006)
rH, preACC, I	Yes	P < T		No	Kasai et al. (2008)
H, ACC, A, mPFC	No	P < T	GM volume loss covaries with P and T	No	O'Doherty et al. (2017)
lH, lACC, mPFC, A, I; A, lI, MTG	No	P < T (single); (prolonged)	lACC loss covaries with CAPS for those with PTSD	Yes	Meng et al. (2016)
lH, calcarine cortex, lPHG	No	P < T	lH, calcarine cortex, lparaH loss covaries with CAPS if have PTSD and small if not PTSD	No	Zhang et al. (2011)
rH, pgACC, lI, rI pgACC	Yes	P < T; P < C		No	Kasai et al. (2008)
lACC,	No	P < T	lACC loss covaries with CAPS in PTSD subjects	No	Yamasue et al. (2003)
ACC	Yes	P < C	ACC loss proportional to 1/CAPS	No	Woodward et al. (2006)
sgACC, lI, lMTG, rMFG, caudate	Yes	T < C; P < C	sgACC, lI, lMTG, caudate loss covaries with CAPS if PTSD and small if not PTSD	No	Herringa et al. (2012)
pgACC	No	P < C		No	Rocha-Rego et al. (2012)
ACC, vmPFC, lMTG, lH	Yes	P < T	ACC, H, MTG, vmPFC loss greater in P versus T	Yes	Kuhn and Gallinat (2012)
A, lPHG	No	P < T		Yes	Del Casale et al. (2022)
lA, rA, lH	Yes	P < T	A, loss does not covary with CAPS	Yes	Morey et al. (2012)
lI, lPHG, lPC	No	P < T	lI, PCC and parahipp loss covaries with trauma load	No	Nardo et al. (2013)
mPFC, lOC; lH, rSFG, lMTC	No	P < C; P < T		Yes	Li et al. (2014)
PFC, OL	No	P < C		No	Tan et al. (2013)

(continued)

Table 10.1 (continued)

Site of grey matter loss	Combat related	Comparative results	Correlation with CAPS/CES	Meta-analysis/review	Authors/sources
vmPFC, dACC	Yes	??	Loss covaries with deployment duration without PTSD but loss areas similar to those with PTSD	No	Butler et al. (2017)
lMFG, rPHG	No	??	rPHG loss covaries with CAPS if have PTSD	No	Azma et al. (2022)
MFG, PCC, PFC, H, ACC, I, A	Yes	P < C/T, P < T	ROI studies	Yes	Siehl et al. (2023)
STG, MFG	Yes	??	Loss covaries with CES (combat exposure therapy) as does CAPS with CES	No	Averill et al. (2017)
lSFG, lCu, rCu, lPa, rPa	Yes	??	Loss covaries with CES but not in those with PTSD	No	Wrocklage et al. (2017)
r pulvinar, l pallidum; lACC, MFL	No	T < C; P < C	lACC loss covaries with CAPS and T if no PTSD. If PTSD in pACC and MFL	No	Chen et al. (2012)
Frontal and occipital lobes	No	P < C	Frontal and occipital lobes loss covaries with CAPS if have PTSD	No	Tavanti et al. (2012)

A amygdala, *ACC* anterior cingulate cortex, *Cu* cuneus, *H* hippocampus, *I* insula, *MFL* middle frontal lobe, *MTG* middle temporal gyrus, *MFG* middle frontal gyrus, *OL* occipital lobe, *P* perirhinal cortex, *Pa* paracentral gyrus, *PCC* posterior cingulate cortex, *PFC* prefrontal cortex, *PHG* parahippocampal gyrus, *SFG* superior frontal gyrus, *STG* superior temporal gyrus, *TL* temporal lobe, *l* left, *r* right, *pg* pregenual, *sg* subgenual, *m* medial, *vm* ventromedial, *d* dorsal. See Tench et al. (2018) for a different conclusion from a meta-analysis. *C* control group (healthy), *CAPS*, clinician-administered PTSD scale, *CES* combat exposure scale, *P* PTSD group, *PTSD* posttraumatic stress disorder, *T* trauma-exposed without PTSD

Where grey matter loss overlaps, there is evidence that the loss is more severe in PTSD patients. This can be seen in comparisons of the hippocampus,[4] ACC[5] and PFC[6] in PTSD patients and trauma-exposed but non-PTSD individuals.

The severity of grey matter changes in particular parts of the brain correlates with duration of trauma exposure (Fig. 10.2) and specific symptom severity (Fig. 10.3).[7]

---

[4] Woon et al. (2010).

[5] Karl et al. (2006).

[6] Meng et al. (2016).

[7] O'Doherty et al. (2017) and Butler et al. (2017).

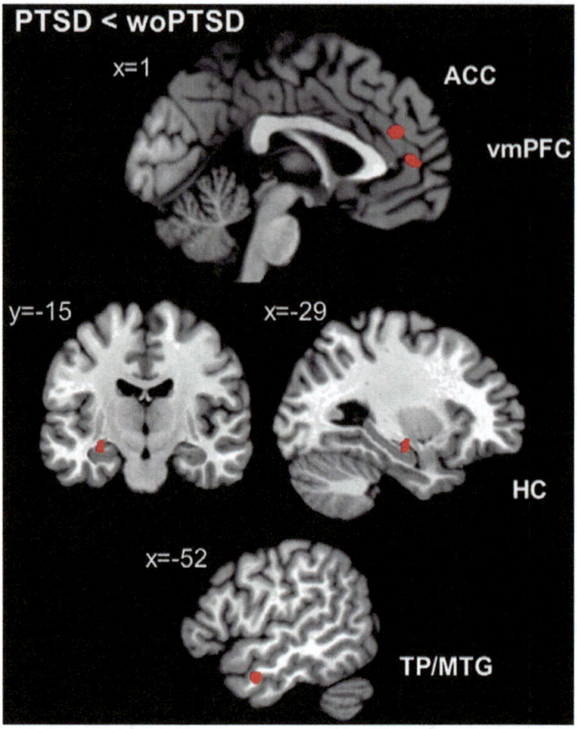

Fig. 10.1 Grey matter loss in victims with posttraumatic stress disorder (PTSD) compared with loss in victims of trauma without PTSD (woPTSD) in the anterior cingulate cortex (ACC) and ventromedial prefrontal cortex (vmPFC), the hippocampus (HC) (front and side views) and the temporal pole and middle temporal gyrus (TP/MTG). (Reproduced from Kühn and Gallinat (2013). Creative Commons CC BY 4.0 (https://creativecommons.org/licenses/by/4.0/))

Collectively, ROI and VBM studies demonstrate both shared and PTSD-specific grey matter loss patterns. While trauma exposure alone leads to atrophy in key hubs (hippocampus, ACC, PFC), PTSD amplifies these changes and extends involvement to regions critical for fear processing and emotional regulation (amygdala, sgACC, insula). Severity correlates with combat exposure length and symptom clusters, underscoring dose–response relationships.

Attempts have been made to relate grey matter loss to the specific PTSD symptoms clusters of PTSD—intrusion (B), avoidance (C), negative alterations in cognition and mood (D) and alterations in arousal and reactivity (E) (see Sect. 2.1)—but with little agreement among the studies (Table 10.2).

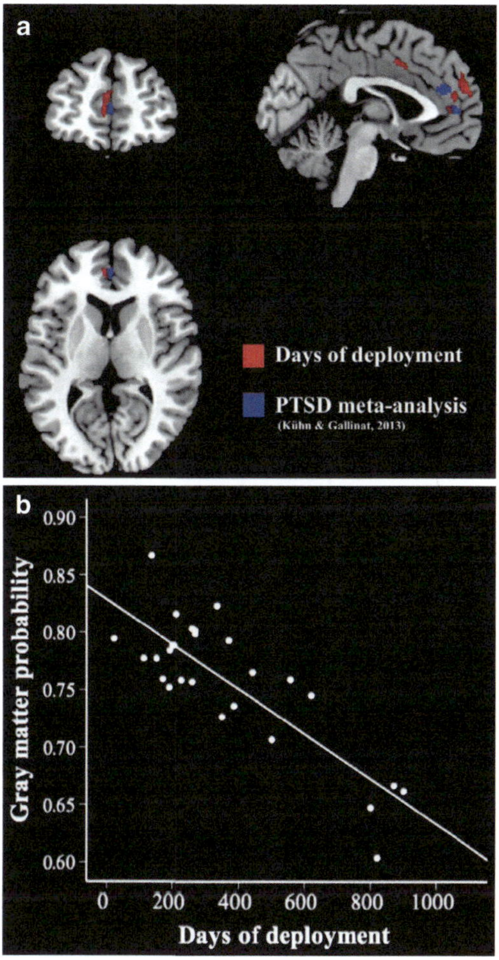

Fig. 10.2 (**a**) Brain regions showing grey matter loss in the left ventromedial prefrontal cortex and in the bilateral dorsal regions mapped to PTSD patients with military deployment and control group from a PTSD meta-analysis. (**b**) Scatterplot depicting the negative correlation between the combined grey matter probability values of the two regions per individual and days of deployment. (Reproduced from Butler et al. (2017). Creative Commons CC BY 4.0 (https://creativecommons.org/licenses/by/4.0/))

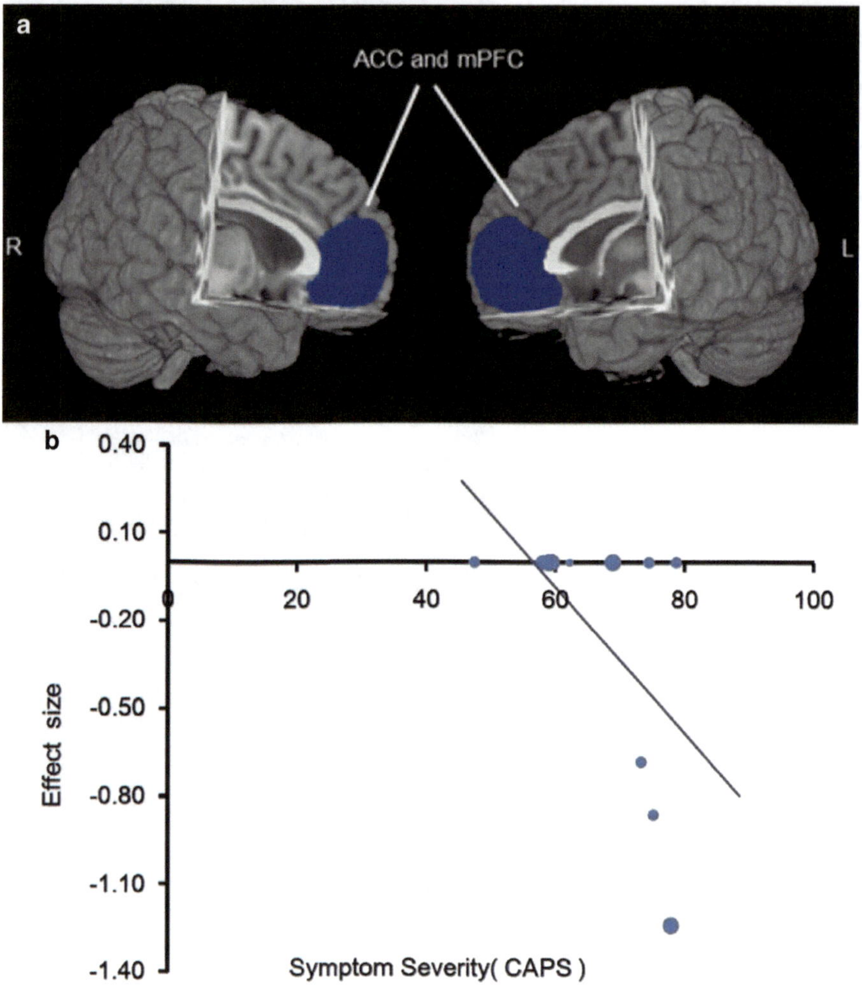

Fig. 10.3 Association between clinician-administered PTSD scale (CAPS) and grey matter (GM) alterations in patients with PTSD. (**a**) The brain areas associating with CAPS scores in PTSD patients, being the anterior cingulate cortex (ACC) and medial prefrontal cortex (mPFC). (**b**) The relationship between CAPS scores and GM alterations of ACC and mPFC in (**a**). (Reproduced from Meng et al. (2016). Creative Commons CC BY 4.0 (https://creativecommons.org/licenses/by/4.0/))

Table 10.2 Findings of studies correlating grey matter loss in specific brain regions to PTSD symptom clusters

Intrusion (A)	Avoidance (B)	Negative alterations in cognition and mood (D)	Changes in arousal and reactivity (E)	Study
r postcentral gyrus; l fusiform gyrus	r amygdala–hippocampal complex; r frontal middle gyrus; l frontal inferior orbital gyrus; r and l cerebellum		r fusiform gyrus; temporal inferior gyrus; l temporal middle gyrus[a]; r superior medial frontal gyrus[a]	Weber et al. (2013)
Middle temporal gyrus; inferior occipital gyrus; operculum/insula				Kroes et al. (2011)
l inferior temporal gyrus[a]; l middle temporal gyrus[a]; r parahippocampal and medial temporal gyrus[a]	l lateral fissure; postcentral gyrus; middle/lateral occipital gyrus; posterior cingulate; superior occipital gyrus; r amygdala		l amygdala[a]	Crombie et al. (2021)
			Amygdala[a]	Hartley et al. (2011)
Ventrolateral PFC[a], superior PFC[a]	l lateral orbitofrontal[a]	l dorsal anterior cingulate cortex[a], l dorsolateral & ventrolateral PFC[a], l inferior parietal cortex[a]	l dorsolateral PFC[a], l superior PFC[a], r lateral occipital cortex[a], r posterior cingulate[a], l lingual gyrus[a], l precuneus[a]	Wrocklage et al. (2017)
R globus pallidus, putamen, l globus pallidus, posterior cingulate gyrus[a], l fusiform gyrus[a]	rI[a], midcingulate gyrus[a], temporal gyrus [a](l inferior, r superior, middle)		r anterior cingulate cortex	Azma et al. (2022)

l left, *r* right, *PFC* prefrontal cortex
[a]Negative correlations

10.2 Grey Matter Loss in the Anterior Cingulate Cortex and PTSD

The ACC is implicated in emotional regulation, cognitive control and fear extinction. Grey matter loss in the ACC in patients with PTSD correlates with several behavioural manifestations that reflect disrupted ACC-mediated integration of cognitive and affective information:

- *Impaired self-regulation*: increased impulsivity, risk-taking and emotional dysregulation due to diminished top–down control
- *Dysregulated fear response*: hyperactivation of the amygdala and insula, impaired extinction learning and inappropriate responses to neutral stimuli
- *Attention and cognitive control deficits*: difficulty sustaining focus, reduced feedback sensitivity and impaired decision-making
- *Default mode network (DMN) dysfunction*: persistence of self-referential processing, intrusive trauma memories and ruminative thought patterns

10.3 Relation Between Grey Matter Loss and Resting State Networks

A revolution in our understanding of brain function was the discovery that patterns of spontaneous correlated activity occur in the brain even at rest using a technique known as functional magnetic resonance imaging (fMRI).[8] Each of these correlated sets of activity constitutes a resting state network (RSN), of which about 20 have been widely recognized.[9] Three RSNs are of special interest in the context of PTSD—the default mode network (DMN), the salience network (SN) and the central executive network (CEN), each defined in terms of their anatomical localization. Malfunction of these networks can give rise to the clinical symptoms of PTSD: altered sense of self with regard to the DMN, executive dysfunction with regard to the CEN and changes in arousal with respect to the SN (Fig. 10.4).[10]

Grey matter loss in PTSD affects the DMN, the SN and the CEN in different ways (Table 10.3). Loss of grey matter in these RSNs also impacts PTSD function (Table 10.4).

---

[8] Biswal et al. (1995).

[9] Yeo et al. (2011).

[10] Lanius et al. (2015).

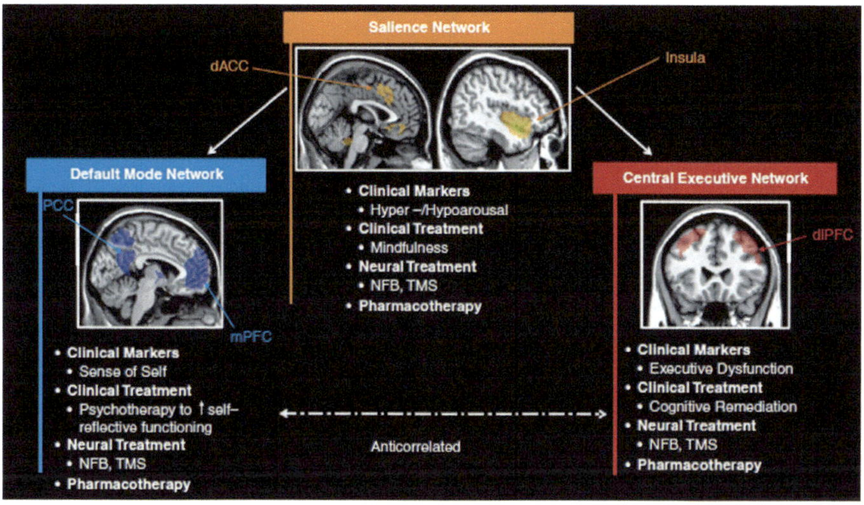

Fig. 10.4 Neuro-scientifically informed treatment interventions in psycho-traumatology: Three intrinsic networks, including the central executive network (CEN), salience network (SN) and default mode network (DMN), may be associated with specific clinical symptoms observed in PTSD, including cognitive dysfunction (CEN), hyper- and hypoarousal/interoception (SN) and an altered sense of self (DMN). Specific testable treatment interventions targeted to restore each one of these brain networks and related clinical dysfunction are suggested. (Reproduced from Lanius et al. (2015). Creative Commons CC BY 4.0 (https://creativecommons.org/licenses/by/4.0/))

Table 10.3 Effect of grey matter (GM) loss on the default mode network (DMN), salience network (SN) and central executive network (CEN)

Network	Core nodes	Structural loss overlap	Observations
DMN	mPFC (vmPFC, pgACC, sgACC) PCC Precuneus IPL Lateral temporal cortex (MTG) HC/PHG	Frequent mPFC, PCC, HC/PHG losses across multiple studies[a]	GM loss in core nodes reflects synaptic and dendritic density, weakening intrinsic connectivity Impaired DMN integrity manifests as disrupted self-referential and autobiographical memory processes, contributing to intrusive recollections and fragmented narratives
SN	Anterior insula dACC Amygdala	Significant atrophy in: Anterior insula[b] ACC[c] Amygdala[d]	Despite GM loss, SN connectivity often shows hyperactivation in PTSD. This paradox could be resolved by compensatory disinhibition due to reduced cortical regulation, neurochemical dysregulation (e.g. increased noradrenergic tone) or maladaptive network reorganization, leading to heightened threat detection and hypervigilance

(continued)

Table 10.3 (continued)

Network	Core nodes	Structural loss overlap	Observations
CEN	dlPFC (MFG, SFG) PPC	Atrophy in dlPFC and parietal regions[e]	Reduced GM in executive hubs undermines working memory, attentional control and goal-directed behaviour Functional hypoconnectivity within the CEN and between CEN–DMN circuits impairs cognitive flexibility and emotion regulation, exacerbating PTSD symptomatology

ACC anterior cingulate cortex, *dACC* dorsal ACC, *dlPFC* dorsolateral PFC, *HC* hippocampus, *IPL* inferior parietal lobule, *mPFC* medial PFC, *MFG* middle frontal gyrus, *MTG* middle temporal gyrus, *PCC* posterior cingulate cortex, *PFC* prefrontal cortex, *pgACC* pregenual ACC, *PHG* parahippocampal gyrus, *PPC* posterior parietal cortex, *PTSD* posttraumatic stress disorder, *SFG* superior frontal gyrus, *sgACC* subgenual ACC, vmPFC, ventromedial PFC

[a]Meng et al. (2016), Li et al. (2014), Herringa et al. (2012) and Nardo et al. (2013)

[b]Meng et al. (2016), Nardo et al. (2013) and Herringa et al. (2012)

[c]Woodward et al. (2006), Butler et al. (2017) and Yamasue et al. (2003)

[d]O'Doherty et al. (2017), Meng et al. (2016) and Del Casale et al. (2022)

[e]Wrocklage et al. (2017), Averill et al. (2017), Tan et al. (2013), Nardo et al. (2013) and Tavanti et al. (2012)

Table 10.4 Impact of grey matter loss in resting state networks on functional changes in posttraumatic stress disorder (PTSD)

Network	Functional change in PTSD	Structural change	Structure–function relation
DMN	Decreased intrinsic connectivity	Widespread GM loss in the medial prefrontal cortex, posterior cingulate cortex, hippocampus and parahippocampal gyrus	Direct match: synaptic loss limits self-referential processing
SN	Increased connectivity (hyperactivity)	Moderate GM loss in the anterior insula, dorsal ACC and amygdala	Inverse match: disinhibition or compensation may yield hyperactivity
CEN	Decreased executive connectivity	GM loss in the dorsolateral prefrontal cortex and posterior parietal cortex	Partial match: structural deficits explain many cognitive control impairments

ACC anterior cingulate cortex, *CEN* central executive network, *DMN* default mode network, *GM* grey matter, *SN* salience network

10.4 Conclusions

The structural degradation in core nodes within the DMN closely maps onto the functional disconnection of the DMN. This is a strong correspondence between loss of grey matter in specific areas of the brain and PTSD symptoms.

Similarly, grey matter loss in core nodes within the CEN reflects reduced synapse connectivity corresponding to some behaviours in PTSD, such as reduced executive controls. However, some studies show more variability in structural findings for the CEN than for DMN or SN.

In contrast, the grey matter loss in SN nodes does not explain the increased functional connectivity. In fact, it presents a paradox, with damaged nodes showing heightened communication. This suggests a decoupling of structure and function in the SN, possibly mediated by network-level dynamics, neurochemical changes (e.g. increased noradrenaline) or compensatory overdrive.

References

Akiki TJ, Averill CL, Abdallah CG (2017) A network-based neurobiological model of PTSD: evidence from structural and functional neuroimaging studies. Curr Psychiatry Rep 19(11):81. https://doi.org/10.1007/s11920-017-0840-4

Averill LA, Abdallah CG, Pietrzak RH, Averill CL, Southwick SM, Krystal JH, Harpaz-Rotem I (2017) Combat exposure severity is associated with reduced cortical thickness in combat veterans: a preliminary report. Chronic Stress 1:2470547017724714. https://doi.org/10.1177/2470547017724714

Azma S, Thompson R, Bermudez D, Renton R, Adeyemo A, Meyerhoff J, Amdur R, Green B, Dutton MA, VanMeter JW (2022) PTSD symptom sub-cluster severity and gray matter volume. medRxiv. https://doi.org/10.1101/2022.07.26.22278078

Biswal B, Yetkin FZ, Haughton VM, Hyde JS (1995) Functional connectivity in the motor cortex of resting human brain using echo-planar MRI. Magn Reson Med 34(4):537–541. https://doi.org/10.1002/mrm.1910340409.

Butler O, Adolf J, Gleich T, Willmund G, Zimmermann P, Lindenberger U, Gallinat J, Kühn S (2017) Military deployment correlates with smaller prefrontal gray matter volume and psychological symptoms in a subclinical population. Transl Psychiatry 7(2):e1031. https://doi.org/10.1038/tp.2016.288

Chen Y, Fu K, Feng C, Tang L, Zhang J, Huan Y, Cui J, Mu Y, Qi S, Xiong L, Ma C, Wang H, Tan Q, Yin H (2012) Different regional gray matter loss in recent onset PTSD and non-PTSD after a single prolonged trauma exposure. PLoS One 7(11):e48298. https://doi.org/10.1371/journal.pone.0048298

Crombie KM, Ross MC, Letkiewicz AM, Sartin-Tarm A, Cisler JM (2021) Differential relationships of PTSD symptom clusters with cortical thickness and grey matter volumes among women with PTSD. Sci Rep 11(1):1825. https://doi.org/10.1038/s41598-020-80776-2

Del Casale A et al (2022) Grey matter volume reductions of the left hippocampus and amygdala in PTSD: a coordinate-based meta-analysis of magnetic resonance imaging studies. Neuropsychobiology 81(4):257–264. https://doi.org/10.1159/000522003

Hartley CA, Fischl B, Phelps EA (2011) Brain structure correlates of individual differences in the acquisition and inhibition of conditioned fear. Cereb Cortex 21(9):1954–1962. https://doi.org/10.1093/cercor/bhq253

Herringa R, Phillips M, Almeida J, Insana S, Germain A (2012) Post-traumatic stress symptoms correlate with smaller subgenual cingulate, caudate, and insula volumes in unmedicated combat veterans. Psychiatry Res Neuroimaging 203(2–3):139–145. https://doi.org/10.1016/j.pscychresns.2012.02.005

Karl A, Schaefer M, Malta LS, Dörfel D, Rohleder N, Werner A (2006) A meta-analysis of structural brain abnormalities in PTSD. Neurosci Biobehav Rev 30(7):1004–1031. https://doi.org/10.1016/j.neubiorev.2006.03.004

Kasai K, Yamasue H, Gilbertson MW, Shenton ME, Rauch SL, Pitman RK (2008) Evidence for acquired pregenual anterior cingulate gray matter loss from a twin study of combat-related posttraumatic stress disorder. Biol Psychiatry 63(6):550–556. https://doi.org/10.1016/j.biopsych.2007.06.022

Kroes MC, Whalley MG, Rugg MD, Brewin CR (2011) Association between flashbacks and structural brain abnormalities in posttraumatic stress disorder. Eur Psychiatry 26(8):525–531. https://doi.org/10.1016/j.eurpsy.2011.03.002

Kühn S, Gallinat J (2013) Gray matter correlates of posttraumatic stress disorder: a quantitative meta-analysis. Biol Psychiatry 73(1):70–74. https://doi.org/10.1016/j.biopsych.2012.06.029

Lanius RA, Frewen PA, Tursich M, Jetly R, McKinnon MC (2015) Restoring large-scale brain networks in PTSD and related disorders: a proposal for neuroscientifically informed treatment interventions. Eur J Psychotraumatol 6:27313. https://doi.org/10.3402/ejpt.v6.27313

Li L et al (2014) Grey matter reduction associated with posttraumatic stress disorder and traumatic stress. Neurosci Biobehav Rev 43:163–172. https://doi.org/10.1016/j.neubiorev.2014.04.003

Meng L et al (2016) Trauma-specific grey matter alterations in PTSD. Sci Rep 6:33748. https://doi.org/10.1038/srep33748

Morey RA et al (2012) Amygdala volume changes in posttraumatic stress disorder in a large case-controlled veterans group. Arch Gen Psychiatry 69(11):1169–1178. https://doi.org/10.1001/archgenpsychiatry.2012.50

Nardo D et al (2013) Gray matter volume alterations related to trait dissociation in PTSD and traumatized controls. Acta Psychiatr Scand 128(3):222–233. https://doi.org/10.1111/acps.12026

O'Doherty DCM, Tickell A, Ryder W, Chan C, Hermens DF, Bennett MR, Lagopoulos J (2017) Frontal and subcortical grey matter reductions in PTSD. Psychiatry Res Neuroimaging 266:1–9. https://doi.org/10.1016/j.pscychresns.2017.05.008

Rocha-Rego V et al (2012) Decreased premotor cortex volume in victims of urban violence with posttraumatic stress disorder. PLoS One 7(8):e42560. https://doi.org/10.1371/journal.pone.0042560

Siehl S, Zohair R, Guldner S, Nees F (2023) Gray matter differences in adults and children with posttraumatic stress disorder: a systematic review and meta-analysis of 113 studies and 11 meta-analyses. J Affect Disord 333:489–516. https://doi.org/10.1016/j.jad.2023.04.028

Tan L et al (2013) Brain structure in post-traumatic stress disorder: a voxel-based morphometry analysis. Neural Regen Res 8(26):2405–2414. https://doi.org/10.3969/j.issn.1673-5374.2013.26.001

Tavanti M et al (2012) Evidence of diffuse damage in frontal and occipital cortex in the brain of patients with post-traumatic stress disorder. Neurol Sci 33(1):59–68. https://doi.org/10.1007/s10072-011-0659-4

Tench CR, Tanasescu R, Jethwa KD, Constantinescu CS (2018) Coordinate-based meta-analysis of whole-brain voxel-based morphometry studies does not show evidence of grey matter loss specific to PTSD. bioRxiv. https://doi.org/10.1101/265496

Weber M et al (2013) Voxel-based morphometric gray matter correlates of posttraumatic stress disorder. J Anxiety Disord 27(4):413–419. https://doi.org/10.1016/j.janxdis.2013.04.004

Woodward SH, Kaloupek DG, Streeter CC, Martinez C, Schaer M, Eliez S (2006) Decreased anterior cingulate volume in combat-related PTSD. Biol Psychiatry 59(7):582–587. https://doi.org/10.1016/j.biopsych.2005.07.033

Woon FL, Sood S, Hedges DW (2010) Hippocampal volume deficits associated with exposure to psychological trauma and posttraumatic stress disorder in adults: a meta-analysis. Prog Neuro-Psychopharmacol Biol Psychiatry 34(7):1181–1188. https://doi.org/10.1016/j.pnpbp.2010.06.016

Wrocklage KM et al (2017) Cortical thickness reduction in combat exposed U.S. veterans with and without PTSD. Eur Neuropsychopharmacol 27(5):515–525. https://doi.org/10.1016/j.euroneuro.2017.02.010

Yamasue H et al (2003) Voxel-based analysis of MRI reveals anterior cingulate gray-matter volume reduction in posttraumatic stress disorder due to terrorism. Proc Natl Acad Sci USA 100(15):9039–9043. https://doi.org/10.1073/pnas.1530467100

Yeo BT, Krienen FM, Sepulcre J, Sabuncu MR, Lashkari D, Hollinshead M, Roffman JL, Smoller JW, Zöllei L, Polimeni JR, Fischl B, Liu H, Buckner RL (2011) The organization of the human cerebral cortex estimated by intrinsic functional connectivity. J Neurophysiol 106(3):1125–1165. https://doi.org/10.1152/jn.00338.2011

Zhang L, Zhang Y, Li L, Li Z, Li W, Ma N, Hou C, Zhang Z, Zhang Z, Wang L, Duan L, Lu G (2011) Different white matter abnormalities between the first-episode, treatment-naive patients with posttraumatic stress disorder and generalized anxiety disorder without comorbid conditions. J Affect Disord 133(1–2):294–299. https://doi.org/10.1016/j.jad.2011.03.040

Chapter 11
Restoring Grey Matter Through Treatment

11.1 Treatments to Ameliorate PTSD

Psychotherapies, pharmacotherapies and neurotherapies used in alleviating PTSD symptoms can lead to restoration of grey matter. Pharmacological treatments (see Chap. 8), such as selective serotonin reuptake inhibitors (SSRIs), have been shown to increase hippocampal volume and improve cognitive function.[1] Ketamine restores grey matter in the inferior frontal gyrus as well as significantly alleviates PTSD.[2]

Psychotherapies that decrease PTSD symptoms (see Chap. 7), such as eye-movement desensitization and reprocessing (EMDR) and trauma-focused cognitive-behavioural therapies (TF-CBT), also restore grey matter in the hippocampus, the dorsolateral and orbitofrontal cortices and the middle frontal gyrus.[3] The anterior cingulate cortex (ACC) and medial prefrontal cortex (mPFC) also benefit from EMDR. TF-CBT produces gains in the dorsolateral and ventrolateral prefrontal cortices, as well as in the entorhinal and cuneus regions.[4] Brief eclectic psychotherapy produces only slight hippocampal change.[5]

Neurotherapies that ameliorate PTSD symptoms (see Chaps. 14 and 15) include neurofeedback, which promotes frontal and temporal cortex thickening.[6]

These observations are summarized in Table 11.1 and illustrated in Fig. 11.1.

---

[1] Vermetten et al. (2003), Bremner and Vermetten (2004), Bossini et al. (2007) and Thomaes et al. (2014).

[2] Dai et al. (2020).

[3] Boukezzi et al. (2017), Letizia et al. (2007), Laugharne et al. (2016), Butler et al. (2017), Nardo et al. (2010), Bossini et al. (2017), Levy-Gigi et al. (2013) and Garrett et al. (2021).

[4] Garrett et al. (2021).

[5] Lindauer et al. (2005).

[6] Ghaziri et al. (2013).

M. Bennett, *Veteran's Stress, Trauma and Synapse Rehabilitation*,
https://doi.org/10.1007/978-981-95-9594-5_11

Table 11.1 The effects on grey matter of different treatments known to ameliorate PTSD symptoms

Treatment	Type of treatment	Restoration effects on grey matter
Eye-movement desensitization and reprocessing	Psychotherapy	Positive changes to the inferior frontal gyrus (dlPFC, OFC), ACC, middle frontal gyrus and mPFC[a] Positive changes to the left and right hippocampus (10% increase)[b,c] Positive changes to the left amygdala[d] Positive changes to the posterior cingulate, anterior insula, right amygdala and parahippocampus[e] Positive changes to the left hippocampus and left thalamus (decrease)[f]
Prolonged exposure	Psychotherapy	No effect on the hippocampus[d]
Trauma-focused cognitive behaviour therapy	Psychotherapy	Positive changes to the left and right hippocampus[g] Positive changes to the right dlPFC, left vlPFC, right entorhinal and left cuneus[h]
Brief eclecric psychotherapy	Psychotherapy	No change to the left or right hippocampus[i]
SSRIs (paroxetine)	Pharmacotherapy	Paroxetine: positive changes to the left and right hippocampus (4–5% increase)[j] Sertraline: positive changes to the left and right hippocampus (10%)[k] Unspecified: positive changes to the right hippocampus (3.7–9.9% increase)[l]
Ketamine	Pharmacotherapy	Positive changes to the inferior frontal gyrus (opercular part)[m]
Neurofeedback	Neurotherapy	Positive changes to the right middle frontal gyrus, inferior temporal gyrus, left superior frontal gyrus and other areas[n]

ACC anterior cingulate cortex, *dlPFC* dorsolateral PFC, *mPFC* medial PFC, *OFC* orbitofrontal cortex, *PFC* prefrontal cortex, *SSRI* selective serotonin reuptake inhibitors, *vlPFC* ventrolateral PFC

[a]Boukezzi et al. (2017)
[b]Letizia et al. (2007)
[c]Butler et al. (2017)
[d]Laugharne et al. (2016)
[e]Nardo et al. (2010)
[f]Bossini et al. (2017)
[g]Levy-Gigi et al. (2013)
[h]Garrett et al. (2021)
[i]Lindauer et al. (2005), see also Levy-Gigi et al. (2013)
[j]Vermetten et al. (2003) and Bremner and Vermetten (2004)
[k]Bossini et al. (2007)
[l]Thomaes et al. (2014)
[m]Dai et al. (2020)
[n]Ghaziri et al. (2013)

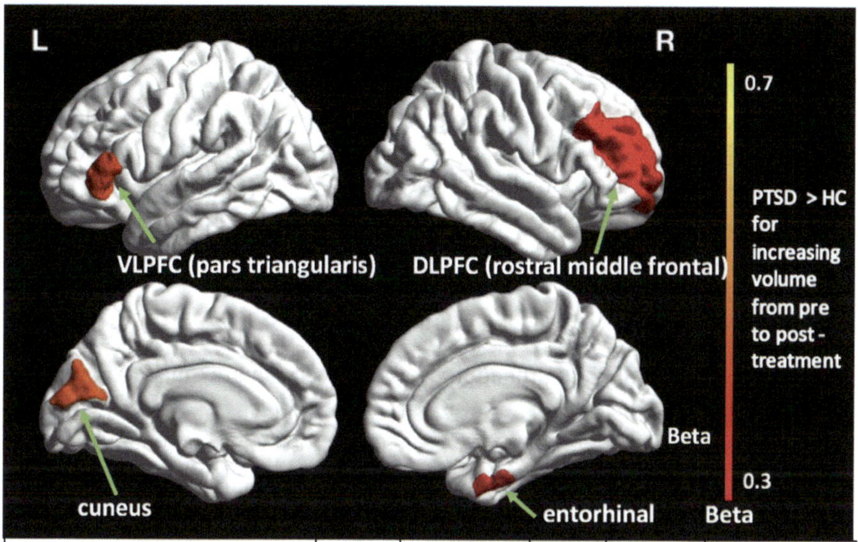

Region	β	SE	df	F	p
R entorhinal	0.330	0.171	2	3.72	.039
R rostral middle frontal (DLPFC)	0.314	0.160	2	3.86	.035
L cuneus	0.443	0.173	3	6.58	.002
L pars triangularis (VLPFC)	0.414	0.153	3	7.30	.001
Age	0.252	0.119	3	4.50	.012

Fig. 11.1 Model of pre- to post-treatment changes, following trauma-focused cognitive-behavioural therapy (TF-CBT), predicting posttraumatic stress disorder (PTSD) versus healthy control group (HC). Changes in the brain volume were calculated as pre- minus post-treatment. The brain images illustrate the location of each region that significantly contributed to the model of longitudinal changes discriminating between individuals in the PTSD and HC groups, with left hemisphere regions on the left side of the image. Regions are colour coded according to standardized beta value, with positive (red) values indicating regions where the PTSD group had relatively greater increases in volume from pre- to post-treatment compared with participants in the HC group. For all regions, the volumes among PTSD group participants increased slightly from pre- to post-treatment, whereas the volumes among those in the HC group decreased slightly. Details of the model are provided in the table below the image. VLPFC, ventrolateral prefrontal cortex; DLPFC, dorsolateral prefrontal cortex. (Reproduced from Garrett et al. (2021) with permission from Wiley)

Table 11.1 shows that the restoration of grey matter in PTSD patients through the use of psychotherapies and pharmacotherapies rehabilitates core regions of resting state networks (see Chap. 10), such as the hippocampus in the default mode network (EMDR; SSRIs) and the dorsolateral prefrontal cortex (dlPFC) in the central executive network (EMDR; TF-CBT).

11.2 Treatments to Ameliorate Suicidality and Depression

Suicidality and suicide attempts, like PTSD, are marked by grey matter loss in brain regions associated with emotion regulation, impulse control and self-referential processing. Particular areas of the brain found to be associated with grey matter loss in suicide attempts are the following:

- dlPFC, orbitofrontal cortex; ACC, anterior cingulate cortex; STC, superior temporal cortex; and POC, postcentral cortex[7]
- Left angular gyrus[8]
- Prefrontal cortex (PFC) and insula[9]

Particular areas of the brain found to be associated with grey matter loss in suicidality are the following:

- Putamen[10]
- Frontal, parietal and temporal regions, as well as the insula and midbrain lentiform nucleus[11]
- PFC, orbitofrontal cortex, left angular gyrus, raphe nucleus, lentiform nucleus, insula, hippocampus, caudate and rectus gyrus[12]
- PFC, ACC, STG and superior temporal gyrus[13]
- Frontal, striatum, insula and parietal regions[14]
- Orbitofrontal cortex and dlPFC[15]
- Orbitofrontal cortex[16]

These studies show that the dlPFC, orbitofrontal cortex and ACC are frequently implicated in individuals exhibiting suicidal ideation or behaviour. *However, at least one study found no subcortical brain structures are affected in patients with suicidality.[17] Particular regions are found in the significant resting state networks: the mPFC, hippocampus and angular gyrus in the default mode network; the ACC and the insula in the salience network; and the dlPFC and the striatum in the central executive network.

Table 11.2 summarizes findings from numerous studies evaluating therapeutic interventions aimed at reversing this structural deterioration.

---

[7] Benedetti et al. (2011).

[8] Lee et al. (2016).

[9] Rizk et al. (2019).

[10] Dombrovski et al. (2012).

[11] Hwang et al. (2010).

[12] Balcioglu and Kose (2018).

[13] Bani-Fatemi et al. (2018).

[14] Kislov et al. (2023).

[15] van Heeringen et al. (2011).

[16] Monkul et al. (2007).

[17] Rentería et al. (2017).

Table 11.2 The effects on grey matter of treatments to ameliorate suicidality and depression

Intervention	Type of treatment	Effect on grey matter	Improved symptoms
Lithium	Pharmacotherapy	Increased grey matter[a]	Bipolar depression
Ketamine	Pharmacotherapy	Increased grey matter in: Caudate, putamen, nucleus accumbens, periaqueductal grey and thalamus[b] Left and right hippocampus[c]	Suicidality
Dialectic behaviour therapy	Psychotherapy	Increased grey matter in ACC, IFG, STG, angular gyrus and supramarginal gyrus[d]	Borderline depression

ACC anterior cingulate cortex, *IFG* inferior frontal gyrus, *STG* superior temporal gyrus
[a]Benedetti et al. (2011)
[b]Gallay et al. (2021)
[c]Dutton et al. (2024)
[d]Mancke et al. (2018)

Table 11.2 shows that treatment with ketamine, dialectic behaviour therapy and sometimes lithium restores grey matter and, it is argued, concomitantly the synapses in these areas (see Chap. 12) while leading to the amelioration of suicidality and to borderline and bipolar depression. It is of considerable interest that ketamine restores grey matter when used to treat both PTSD and suicidality.

References

Balcioglu Y, Kose S (2018) Neural substrates of suicide and suicidal behaviour: from a neuroimaging perspective. Psychiatry Clin Psychopharmacol 28(3):314–328

Bani-Fatemi A, Tasmim S, Graff-Guerrero A, Gerretsen P, Strauss J, Kolla N, Spalletta G, De Luca V (2018) Structural and functional alterations of the suicidal brain: an updated review of neuroimaging studies. Psychiatry Res Neuroimaging 278:77–91. https://doi.org/10.1016/j.pscychresns.2018.05.008

Benedetti F, Radaelli D, Poletti S, Locatelli C, Falini A, Colombo C, Smeraldi E (2011) Opposite effects of suicidality and lithium on gray matter volumes in bipolar depression. J Affect Disord 135(1–3):139–147. https://doi.org/10.1016/j.jad.2011.07.006

Bossini L et al (2007) Changes in hippocampal volume in patients with post-traumatic stress disorder after sertraline treatment. J Clin Psychopharmacol 27(2):233–235. https://doi.org/10.1097/01.jcp.0000264979.09862.a8

Bossini L et al (2017) Morphovolumetric changes after EMDR treatment in drug-naïve PTSD patients. Riv Psichiatr 52(1):24–31. https://doi.org/10.1708/2631.27051

Boukezzi S et al (2017) Grey matter density changes of structures involved in posttraumatic stress disorder (PTSD) after recovery following eye movement desensitization and reprocessing (EMDR) therapy. Psychiatry Res Neuroimaging 266:146–152. https://doi.org/10.1016/j.pscychresns.2017.06.009

Bremner JD, Vermetten E (2004) Neuroanatomical changes associated with pharmacotherapy in posttraumatic stress disorder. Ann N Y Acad Sci 1032:154–157. https://doi.org/10.1196/annals.1314.012

Butler O et al (2017) Military deployment correlates with smaller prefrontal gray matter volume and psychological symptoms in a subclinical population. Transl Psychiatry 7(2):e1031. https://doi.org/10.1038/tp.2016.288

Dai D et al (2020) Ketamine normalizes the structural alterations of inferior frontal gyrus in depression. Chronic Stress 4:2470547020980681. https://doi.org/10.1177/2470547020980681

Dombrovski AY et al (2012) The temptation of suicide: striatal gray matter, discounting of delayed rewards, and suicide attempts in late-life depression. Psychol Med 42(6):1203–1215. https://doi.org/10.1017/S0033291711002133

Dutton M et al (2024) Hippocampal subfield volumes predict treatment response to oral ketamine in people with suicidality. J Psychiatr Res 169:192–200. https://doi.org/10.1016/j.jpsychires.2023.11.040

Gallay CC et al (2021) Six-week oral ketamine treatment for chronic suicidality is associated with increased grey matter volume. Psychiatry Res Neuroimaging 317:111369. https://doi.org/10.1016/j.pscychresns.2021.111369

Garrett AS et al (2021) Changes in brain volume associated with trauma-focused cognitive behavioral therapy among youth with posttraumatic stress disorder. J Trauma Stress 34(4):744–756. https://doi.org/10.1002/jts.22678

Ghaziri J et al (2013) Neurofeedback training induces changes in white and gray matter. Clin EEG Neurosci 44(4):265–272. https://doi.org/10.1177/1550059413476031

Hwang JP et al (2010) Cortical and subcortical abnormalities in late-onset depression with history of suicide attempts investigated with MRI and voxel-based morphometry. J Geriatr Psychiatry Neurol 23(3):171–184. https://doi.org/10.1177/0891988710363713

Kislov MA et al (2023) Morphofunctional cerebral changes associated with development of suicidal behavior. Sud Med Ekspert 66(4):67–72. https://doi.org/10.17116/sudmed20236604167

Laugharne J et al (2016) Amygdala volumetric change following psychotherapy for posttraumatic stress disorder. J Neuropsychiatry Clin Neurosci 28(4):312–318. https://doi.org/10.1176/appi.neuropsych.16010006

Lee YJ et al (2016) Decreased regional gray matter volume in suicide attempters compared to suicide non-attempters with major depressive disorders. Compr Psychiatry 67:59–65. https://doi.org/10.1016/j.comppsych.2016.02.013

Letizia B, Andrea F, Paolo C (2007) Neuroanatomical changes after eye movement desensitization and reprocessing (EMDR) treatment in posttraumatic stress disorder. J Neuropsychiatry Clin Neurosci 19(4):475–476. https://doi.org/10.1176/jnp.2007.19.4.475

Levy-Gigi E, Szabó C, Kelemen O, Kéri S (2013) Association among clinical response, hippocampal volume, and FKBP5 gene expression in individuals with posttraumatic stress disorder receiving cognitive behavioral therapy. Biol Psychiatry 74(11):793–800. https://doi.org/10.1016/j.biopsych.2013.05.017

Lindauer RJL, Vlieger E-J, Jalink M, Olff M, Carlier IVE, Majoie CBLM, Den Heeten GJ, Gersons BPR (2005) Effects of psychotherapy on hippocampal volume in out-patients with post-traumatic stress disorder: an MRI investigation. Psychol Med 35(10):1421–1431. https://doi.org/10.1017/s0033291705005246

Mancke F et al (2018) Assessing the marks of change: how psychotherapy alters the brain structure in women with borderline personality disorder. J Psychiatry Neurosci 43(3):171–181. https://doi.org/10.1503/jpn.170132

Monkul ES, Hatch JP, Nicoletti MA, Spence S, Brambilla P, Lacerda AL, Sassi RB, Mallinger AG, Keshavan MS, Soares JC (2007) Fronto-limbic brain structures in suicidal and non-suicidal female patients with major depressive disorder. Mol Psychiatry 12(4):360–366. https://doi.org/10.1038/sj.mp.4001919

Nardo D et al (2010) Gray matter density in limbic and paralimbic cortices is associated with trauma load and EMDR outcome in PTSD patients. J Psychiatr Res 44(7):477–485. https://doi.org/10.1016/j.jpsychires.2009.10.014

Rentería ME, Schmaal L, Hibar DP et al (2017) Subcortical brain structure and suicidal behaviour in major depressive disorder: a meta-analysis from the ENIGMA-MDD working group. Transl Psychiatry 7(5):e1116. https://doi.org/10.1038/tp.2017.84

Rizk MM et al (2019) Gray matter volumetric study of major depression and suicidal behavior. Psychiatry Res Neuroimaging 283:16–23. https://doi.org/10.1016/j.pscychresns.2018.11.007

Thomaes K, Dorrepaal E, Draijer N, Jansma EP, Veltman DJ, van Balkom AJ (2014) Can pharmacological and psychological treatment change brain structure and function in PTSD? A systematic review. J Psychiatr Res 50:1–15. https://doi.org/10.1016/j.jpsychres.2013.11.002

van Heeringen C, Bijttebier S, Godfrin K (2011) Suicidal brains: a review of functional and structural brain studies in association with suicidal behaviour. Neurosci Biobehav Rev 35(3):688–698. https://doi.org/10.1016/j.neubiorev.2010.08.007

Vermetten E, Vythilingam M, Southwick SM, Charney DS, Bremner JD (2003) Long-term treatment with paroxetine increases verbal declarative memory and hippocampal volume in posttraumatic stress disorder. Biol Psychiatry 54(7):693–702. https://doi.org/10.1016/s0006-3223(03)00634-6

Chapter 12
Grey Matter Volume Changes and Synaptic Alterations

12.1 Relationships Between Psychiatric Diseases and Changes in Dendritic Volume

About 26% of the volume of grey matter in the brain consists of dendrites.[1] Preclinical studies show that there is a linear relationship between the loss of grey matter volume in the brain following stress and the loss of total dendritic volume (including in the anterior cingulate cortex (ACC) and hippocampus).[2] The dendrites of neurons in the nucleus laminaris have an individual branch length that is determined by the number of synapses on the dendrite, so that the loss of synapses on a particular dendrite through deafferentation leads to shortening of that dendrite without affecting the other dendrites of the neuron.[3] 'Functional' deafferentation of a dendrite, for example, by blocking glutamate receptors, or functional excitation by electrical stimulation of synapses on the dendrite, leads to decreases and increases, respectively, in the total length of the dendrite without affecting adjoining dendrites.[4] Given these observations, the loss of grey matter in particular regions of the brain in individuals suffering from psychiatric diseases, such as posttraumatic stress disorder (PTSD), is due to the loss of neuronal dendrites in these regions, consequent on the loss of synapses. Reconstituting these synapses, and thus the dendrites on which they impinge, thereby re-establishing normal grey matter volume, is therefore essential if remedial treatment for PTSD is to be successful. To what extent is this hypothesis supported by recent observations?

There is considerable evidence showing that grey matter loss in psychiatric diseases is accompanied by the loss of synapses. The hippocampus, ACC and

[1] Kassem et al. (2013).

[2] Kassem et al. (2013).

[3] Sorensen and Rubel (2006).

[4] Sorensen and Rubel (2006).

M. Bennett, *Veteran's Stress, Trauma and Synapse Rehabilitation*,
https://doi.org/10.1007/978-981-95-9594-5_12

dorsolateral prefrontal cortex (dlPFC) are principal sites in the brain for which there is extensive evidence for loss of grey matter in PTSD patients (see Chap. 10). It is then of considerable interest that this loss is accompanied by extensive loss of synapses in these areas, as determined by their labelling with synaptic vesicle protein 2 (SV2),[5] suggesting that they are related. In preclinical observations, the loss of synaptic spines in the auditory cortex following auditory fear conditioning is correlated with the loss of grey matter in this cortex.[6] These observations are summarized in Table 12.1.

Table 12.1 Evidence that volume changes in grey matter reflect volume changes in the extent of dendrites conditional on their synapse numbers

Condition	Loss of grey matter/functional connections (L)	Loss of synapses (L)	Dendritic length shortened (L)	Clinical (C) or preclinical (P) study
MDD/PTSD[a]	SV2A density proportional to functional connectivity	(L) SV2A-labelled synapses Decrease in dlPFC with MDD/PTSD increases		C (dlPFC)
MDD[b]		(L) Synaptic proteins in dlPFC	(L) Dendrites	C (dlPFC)
Worry, anxiety, fatigue[c]		(L) SV2-labelled synapses linearly decline with intensity of worry, anxiety and fatigue		C (dlPFC)
Stressed[d]		(L) Grey matter density directly proportional to synaptic spine density with no changes in cells		P, using auditory fear conditioning (auditory cortex)
Stressed[e]	L of grey matter quantitatively proportional to L of dendrites		(L) Dendrites	P (ACC, hippocampus, amygdala)
Stressed[f]		(L) Spines and dendrites		P (ldACC)
Deafferented[g]		(L) of only deafferented dendrites		P (nucleus laminaris)

(continued)

[5] Holmes et al. (2019).

[6] Keifer Jr et al. (2015).

Table 12.1 (continued)

Condition	Loss of grey matter/functional connections (L)	Loss of synapses (L)	Dendritic length shortened (L)	Clinical (C) or preclinical (P) study
Deafferented[h]		(L) of synapses simultaneous with loss of dendritic complexity		P (retina)

[a]Holmes et al. (2019)
[b]Kang et al. (2012)
[c]Asch et al. (2022)
[d]Keifer Jr et al. (2015)
[e]Kassem et al. (2013)
[f]Chen et al. (2018)
[g]Sorensen and Rubel (2006)
[h]Berry et al. (2015)

12.2 Effectiveness of Ketamine to Restore Synapses

Patients with major depressive disorder (MDD) show a significant loss of synapses in the dlPFC, which accompanies loss of grey matter.[7] Thus, while the loss of grey matter is accompanied by loss of synapses, is there evidence that this correlation is mediated by the loss of dendrites? The decrease in dendritic complexity in the subiculum of patients with psychiatric diseases is associated with a decline in synaptic spine density.[8] A study into ketamine restoration of synapses lost following stress found that 'the diminished volume of the hippocampus in depression … may be critically determined by a loss in neuropil including dendritic branching, dendritic spine complexity, and glial processes'.[9]

Given these observations, it would be expected that the essential requirement for successful PTSD treatment involves the re-establishment of lost synapses. In preclinical experiments, synapse loss in the hippocampus was restored with fluoxetine, a selective serotonin reuptake inhibitor (SSRI).[10] Indeed, the intensity of disorders, such as worry, fatigue and anxiety, are directly correlated with decrease in the extent of SV2-labelled synapses in the dlPFC (Fig. 12.1).[11] The spectacular improvement in the amelioration of psychiatric conditions with ketamine as well as the use of SSRIs is indeed accompanied by the restoration of synapses (Fig. 12.2).[12] Ketamine restores functional connectivity in the dlPFC,

[7] Kang et al. (2012).
[8] Rosoklija et al. (2000).
[9] Stockmeier et al. (2004).
[10] Herrmann et al. (2012).
[11] Asch et al. (2022).
[12] Asch et al. (2024).

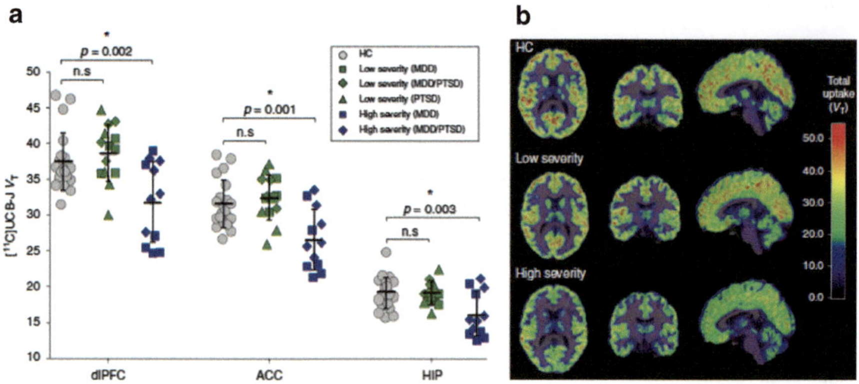

Fig. 12.1 Lower SV2A density in individuals with high-severity depressive symptoms (MDD) vs. HC (healthy control) subjects. (**a**) [11C]UCB-J VT (radioligand [11C]UCB-J binding to the synaptic glycoprotein SV2A and so a synaptic marker) in the dlPFC, ACC and hippocampus across groups (defined below). The low-severity group (in green; $n = 14$) consisted of participants with depression HAMD-17 MDD scores <14, and the high-severity group (in blue; $n = 12$) consisted of participants with HAMD-17 scores ≥14. Clinical subgroups are represented by different symbols (see box insert): the low-severity group consisted of MDD ($n = 5$), PTSD ($n = 5$) and MDD/PTSD ($n = 4$) subjects; the high-severity group consisted of MDD ($n = 6$) and MDD/PTSD ($n = 6$) subjects. Error bars represent standard deviation. (**b**) Representative axial, coronal and sagittal parametric images of [11C] UCB-J VT PET scans from a HC participant (top), low-severity subject (middle) and high-severity subject (bottom). Colour bar represents VT. dlPFC, dorsolateral prefrontal cortex; ACC, anterior cingulate cortex. (Reproduced Holmes et al. (2019). Creative Commons CC BY 4.0 (https://creativecommons.org/licenses/by/4.0/))

insula and caudate of MDD patients, thereby ameliorates MDD in proportion to the restored connectivity.[13] Thus, ketamine serves as a prototypical agent demonstrating the principle that synapse restoration underlies both anatomical and clinical recovery in psychiatric disorders. This reinforces the paradigm that therapeutic strategies must focus on synaptogenesis and dendritic rehabilitation to achieve lasting symptom relief and neural recovery. These observations are summarized in Table 12.2.

[13] Abdallah et al. (2017).

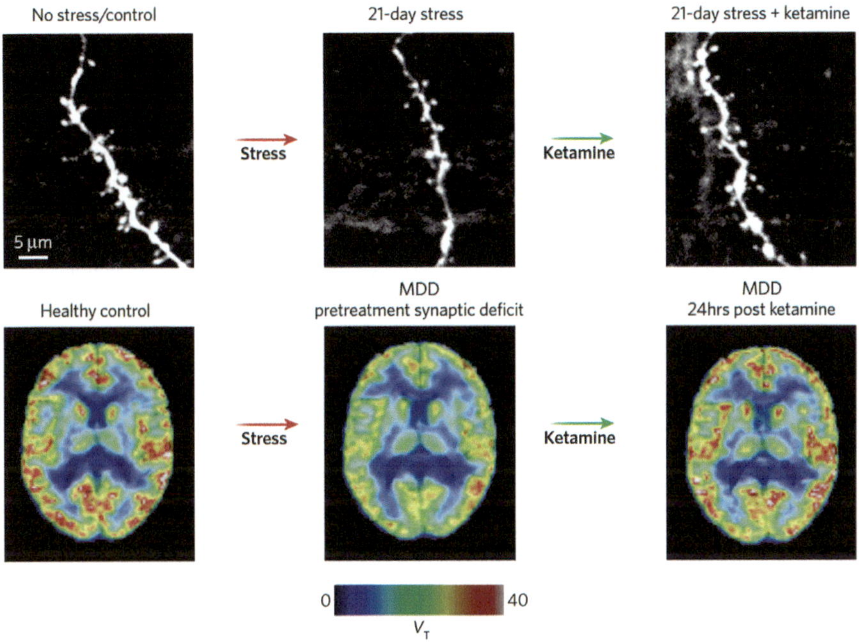

Fig. 12.2 Effect of ketamine treatment following stress-induced loss of synapses. Top row—two-photon microscopy images of neurobiotin-labelled layer V pyramidal cells demonstrating dendrite/synapses that are typical (left), under stress (middle) and post-ketamine treatment (right). Bottom row—representative individual PET images of synaptic density in the human brain with [11C]UCB-J VT (defined in the legend to Fig. 12.1). Healthy adult (left); an age- and sex-matched individual during a depressive episode (middle) and the same individual post-ketamine administration (right). Chronic stress induces alteration in the brain, including changes in synaptic proteins and synaptic density (all rows, middle panel). However, administration of some drugs, with ketamine leading the literature in predominance, can lead to measurable increases in synaptic density. (Reproduced from Asch et al. (2024). Creative Commons CC BY 4.0 (https://creativecommons.org/licenses/by/4.0/))

Table 12.2 Evidence that ketamine and SSRIs restore synapses and ameliorate PTSD/MDD

Treatment	Improved clinical symptom (yes or no)	Restoration (R) of grey matter/ functional connections	Restoration (R) of synapses	Mechanism of restoration	Clinical (C) or preclinical (P) study
SSRI[a] (fluoxetine)	PTSD, yes		(R) Synapse proteins		P (hippocampus)
SSRI[b]	PTSD, yes		(R) Synapses	Principally by increase of BDNF	P (review)

(continued)

Table 12.2 (continued)

Treatment	Improved clinical symptom (yes or no)	Restoration (R) of grey matter/ functional connections	Restoration (R) of synapses	Mechanism of restoration	Clinical (C) or preclinical (P) study
Ketamine[c]	MDD, yes	(R) Functional connections			C (PFC, cingulate, precuneus)
Ketamine, SSRI (escitalopram)[d]	MDD/ PTSD, yes		(R) SV2A-labelled synapses increase in PFC		C (PFC) Review

[a]Herrmann et al. (2012)
[b]Krystal et al. (2017)
[c]Abdallah et al. (2017)
[d]Asch et al. (2024)

References

Abdallah CG, Averill LA, Collins KA, Geha P, Schwartz J, Averill C, DeWilde KE, Wong E, Anticevic A, Tang CY, Iosifescu DV, Charney DS, Murrough JW (2017) Ketamine treatment and global brain connectivity in major depression. Neuropsychopharmacology 42(6):1210–1219. https://doi.org/10.1038/npp.2016.186

Asch RH, Holmes SE, Jastreboff AM, Potenza MN, Baldassarri SR, Carson RE, Pietrzak RH, Esterlis I (2022) Lower synaptic density is associated with psychiatric and cognitive alterations in obesity. Neuropsychopharmacology 47(2):543–522. https://doi.org/10.1038/s41386-021-01111-5

Asch RH, Abdallah CG, Carson RE, Esterlis I (2024) Challenges and rewards of in vivo synaptic density imaging, and its application to the study of depression. Neuropsychopharmacology 50(1):153–163. https://doi.org/10.1038/s41386-024-01913-3

Berry RH, Qu J, John SWM, Howell GR, Jakobs TC (2015) Synapse loss and dendrite remodeling in a mouse model of glaucoma. PLoS One 10(12):e0144341. https://doi.org/10.1371/journal.pone.0144341

Chen C-C, Lu J, Yang R, Ding JB, Zuo Y (2018) Selective activation of parvalbumin interneurons prevents stress-induced synapse loss and perceptual defects. Mol Psychiatry 23(7):1614–1625. https://doi.org/10.1038/mp.2017.159

Herrmann L, Ionescu IA, Henes K, Golub Y, Wang NXR, Buell DR, Holsboer F, Wotjak CT, Schmidt U (2012) Long-lasting hippocampal synaptic protein loss in a mouse model of posttraumatic stress disorder. PLoS One 7(8):e42603. https://doi.org/10.1371/journal.pone.0042603

Holmes SE, Scheinost D, Finnema SJ, Naganawa M, Davis MT, DellaGioia N, Nabulsi N, Matuskey D, Angarita GA, Pietrzak RH, Duman RS, Sanacora G, Krystal JH, Carson RE, Esterlis I (2019) Synaptic density is associated with depression severity and network alterations. Nat Commun 10(1):1529. https://doi.org/10.1038/s41467-019-09562-7

Kang HJ, Voleti B, Hajszan T, Rajkowska G, Stockmeier CA, Licznerski P, Lepack A, Majik MS, Jeong LS, Banasr M, Son H, Duman RS (2012) Decreased expression of synapse-related genes and loss of synapses in major depressive disorder. Nat Med 18(9):1413–1417. https://doi.org/10.1038/nm.2886

Kassem MS, Lagopoulos J, Stait-Gardner T, Price WS, Chohan TW, Arnold JC, Hatton SN, Bennett MR (2013) Stress-induced grey matter loss determined by MRI is primarily due to loss of dendrites and their synapses. Mol Neurobiol 47(2):645–661. https://doi.org/10.1007/s12035-012-8365-7

Keifer OP Jr, Hurt RC, Gutman DA, Keilholz SD, Gourley SL, Ressler KJ (2015) Voxel-based morphometry predicts shifts in dendritic spine density and morphology with auditory fear conditioning. Nat Commun 6:7582. https://doi.org/10.1038/ncomms8582

Krystal JH, Abdallah CG, Averill LA, Kelmendi B, Harpaz-Rotem I, Sanacora G, Southwick SM, Duman RS (2017) Synaptic loss and the pathophysiology of PTSD: implications for ketamine as a prototype novel therapeutic. Curr Psychiatry Rep 19(10):74. https://doi.org/10.1007/s11920-017-0829-z

Rosoklija G, Toomayan G, Ellis SP, Keilp J, Mann JJ, Latov N, Hays AP, Dwork AJ (2000) Structural abnormalities of subicular dendrites in subjects with schizophrenia and mood disorders: preliminary findings. Arch Gen Psychiatry 57(4):349–356. https://doi.org/10.1001/archpsyc.57.4.349

Sorensen SA, Rubel EW (2006) The level and integrity of synaptic input regulates dendrite structure. J Neurosci 26(5):1539–1550. https://doi.org/10.1523/jneurosci.3807-05.2006

Stockmeier CA, Mahajan GJ, Konick LC, Overholser JC, Jurjus GJ, Meltzer HY, Uylings HB, Friedman L, Rajkowska G (2004) Cellular changes in the postmortem hippocampus in major depression. Biol Psych 56(9):640–650. https://doi.org/10.1016/j.biopsych.2004.08.022

Chapter 13
Cellular and Molecular Mechanisms of Synapse Loss and Replacement in Mood and Stress Disorders: BDNF and Microglia

13.1 Brain-Derived Neurotrophic Factor (BDNF) Molecular Mechanisms in Synapse Plasticity

Brain-derived neurotrophic factor (BDNF) is the central regulator of synaptic plasticity, exerting its effects through activation of its high-affinity tropomyosin receptor kinase B (TrkB) and the engagement of multiple intracellular signalling cascades. Under basal conditions, proper synaptic connectivity in the prefrontal cortex (PFC) and hippocampus is maintained by homeostatic mechanisms that include activity-dependent regulation of synaptic proteins, such as GluA1, PSD-95 and synapsin-1. Disruption of BDNF expression and TrkB–mTORC1 signalling contributes to the synaptic deficits observed in stress and depression, whereas enhancement of BDNF release restores synapse number and function.[1]

13.1.1 TrkB Receptor Activation and Downstream Effectors

Upon binding of BDNF, TrkB undergoes autophosphorylation, initiating several distinct pathways (for definition of acronyms, see the legend to Fig. 13.3)[2]:

1. Phospholipase Cγ (PLCγ) pathway: TrkB-mediated activation of PLCγ elevates intracellular Ca^{2+} levels, which promotes neurotransmitter release presynaptically (Fig. 13.3).
2. Mitogen-activated protein kinase (MAPK)/extracellular signal-regulated kinase (ERK) cascade: Engagement of Ras leads to MAPK phosphorylation of synapsin,

[1] See Deyama and Duman (2020), Fig. 1; Park and Poo (2013), Fig. 2.

[2] Park and Poo (2013), Fig. 2.

M. Bennett, *Veteran's Stress, Trauma and Synapse Rehabilitation*,
https://doi.org/10.1007/978-981-95-9594-5_13

facilitating vesicle mobilization, and to transcriptional activation via retrogradely transported signalling endosomes that induce immediate early genes (e.g. c-FOS and ARC), essential for long-term potentiation (Fig. 13.3).

3. PI3K–Akt–mTORC1 axis: Activation of phosphoinositide 3-kinase (PI3K) and Akt enhances both local and global mRNA translation through mTORC1, increasing synthesis of synaptic proteins, such as GluA1 and PSD-95, thereby strengthening synapses (Fig. 13.1).[3]
4. CaMKII and protein kinase C (PKC) activation: These kinases phosphorylate AMPA receptor subunits, increasing their conductance and synaptic insertion, which potentiates postsynaptic responsiveness (Fig. 13.3).
5. Structural modulation of the actin cytoskeleton: TrkB signalling also activates small Rho GTPases, which orchestrate actin polymerization and thus spine morphogenesis (Fig. 13.1).

Additionally, BDNF induces store-operated Ca^{2+} entry via TRPC channels, enhancing dendritic growth and spine maturation.

Turning to presynaptic vesicle dynamics and cadherin-β–catenin interactions: BDNF–TrkB signalling modulates the presynaptic scaffold through phosphorylation of β-catenin at tyrosine 654. In the absence of TrkB activity, β-catenin remains bound to cadherins at the active zone, tethering synaptic vesicle (SV) clusters. Following BDNF stimulation, β-catenin phosphorylation leads to its dissociation from cadherins and dispersion of SV clusters along the axon. TrkB internalization and β-catenin dephosphorylation allow reclustering of SVs, although with an increased rate of small cluster splitting, hypothesized to contribute to the formation of new presynaptic sites and increased synapse density.[4]

13.1.2 Activity-Dependent Versus Constitutive BDNF Release in Antidepressant Responses

Chronic stress reduces BDNF expression and elevates the negative mTORC1 regulator REDD1, resulting in spine loss and impaired synaptic function. Typical monoamine reuptake inhibitors slowly upregulate BDNF expression and result in constitutive BDNF release over weeks, contributing to gradual synaptic remodelling. By contrast, the NMDA receptor antagonist ketamine induces a rapid glutamate surge that triggers activity-dependent release of BDNF, promptly activating TrkB–mTORC1 signalling and translating synaptic proteins to restore spine density and synaptic efficacy within hours (Fig. 13.1) (see Sect. 13.2).[5]

---

[3] Deyama and Duman (2020), Fig. 1.

[4] Bamji et al. (2006) and Chen et al. (2017).

[5] Deyama and Duman (2020).

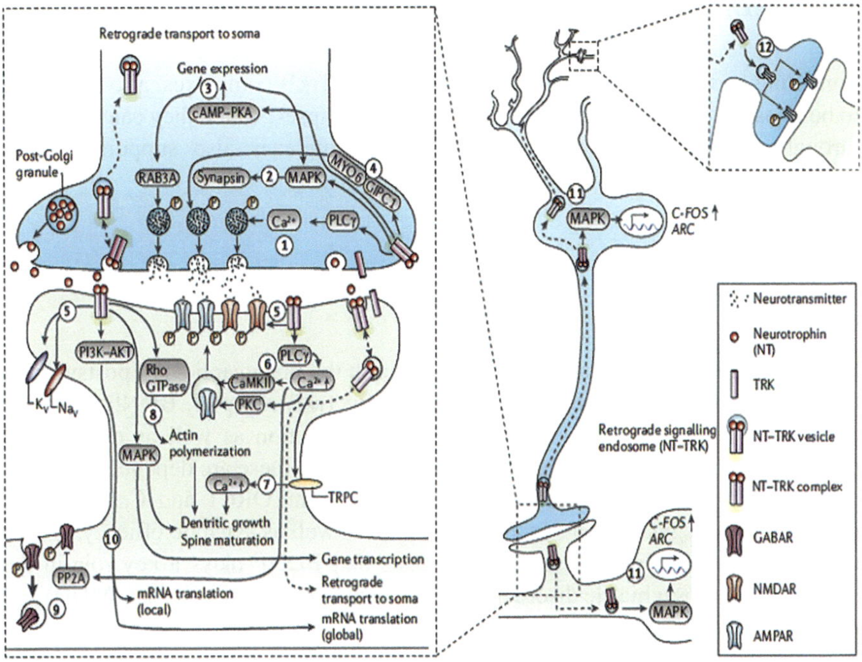

H_y_u_n_g_j_u_ _P_a_r_k_ _a_n_d_ _M_u_-m_i_n_g_ _P_o_o_ _ (2013)

Fig. 13.1 Modulation of synaptic functions by neurotrophins. Neurotrophins (NTs) have diverse pre- and postsynaptic modulatory effects on mature glutamatergic synapses. On the presynaptic side, tropomyosin-related kinase (TRK) signalling could activate the phospholipase Cγ _(PLCγ) pathway to increase the intracellular Ca2+ concentration ([Ca2+]i) (1), mitogen-activated protein kinase (MAPK) phosphorylation of synapsin (2) and RAB3 expression (3), leading to increased presynaptic neurotransmitter release. TRKB-mediated activation of the motor myosin VI (MYO6)–adaptor protein GIPC1 complex also contributes to the increased presynaptic release (4). On the postsynaptic side, TRK signalling causes tyrosine phosphorylation of voltage-gated Na + (NaV) and voltage-gated K+ (KV) channels and NMDA-type glutamate receptor (NMDAR) subunits, modifying the conductance of all these channels (5). TRK signalling also induces activation of Ca2+/calmodulin-dependent kinase II (CaMKII) and protein kinase C (PKC), which phosphorylate AMPA-type glutamate receptor (AMPAR) subunits and increase their synaptic delivery (6). BDNF-induced store-operated activation of transient receptor potential cation channel subfamily C (TRPC) leads to local [Ca2+]i increase, promoting dendritic growth and spine maturation (7). TRK signalling also activates small Rho GTPases that modulate actin cytoskeleton (8). Postsynaptic GABAergic synapses are suppressed by TRK-dependent activation of protein phosphatase 2A (PP2A) that reduces the surface expression of type A GABA receptors (GABAARs) in the dendrite (9). TRK-induced phosphoinositide 3-kinase (PI3K)–AKT signalling enhances both local and global mRNA translation (10). Binding of secreted NTs to their appropriate TRKs leads to endocytic uptake of NT–TRK complexes into 'signalling endosomes', which might be transported retrogradely to the soma of both pre- and postsynaptic neurons, triggering MAPK-dependent transcription of activity-related genes, such as *C-FOS* (also known as *FOS*) and activity-regulated cytoskeleton-associated protein (*ARC*), and genes required for the establishment of long-term synaptic plasticity (11). Long-range transport of signalling endosomes from the axon terminal to the dendrite could potentiate upstream synapses by increasing postsynaptic AMPAR density and conductance (12). Note that secretion of NT-containing post-Golgi granules is shown only at the presynaptic nerve terminal, although release of NTs from NT–TRK endosomes from either pre- or postsynaptic cells may also occur, especially at developing synapses. Dashed arrows indicate pathways that remain to be confirmed. cAMP, cyclic AMP. (Reproduced from Park and Poo (2013) with permission from Springer Nature)

13.1.3 Integration and Long-Term Maintenance

Signalling endosomes internalize secreted BDNF–TrkB complexes, allowing them to be transported to the neuron soma or to other synapses, which they can potentiate through enhancing AMPA receptors. These complexes also support MAPK-dependent transcription of plasticity-related genes (Fig. 13.1).[6]

13.1.4 Conclusion

BDNF has a remarkable range of effects on both the presynaptic and postsynaptic components of the synapse. On activation of its TrkB receptor, BDNF alters the release of transmitters and modulates gene transcription as well as cytoskeletal dynamics and transmitter receptor conductance. All of these are dependent on activated TrkB working on PLCγ, MAPK, PI3K–Akt–mTORC1 and CaMKII/PKC pathways. The regulation of synaptic plasticity, as well as synaptic efficacy, is then critically dependent on BDNF. This suggests that BDNF plays a key role in the amelioration of psychiatric disorders, such as major depressive disorder (MDD) and posttraumatic stress disorder (PTSD).

13.2 Brain-Derived Neurotrophic Factor (BDNF) and Synapse Growth/Replacement

13.2.1 Ketamine Increases BDNF and Thus Restores Synapses

Preclinical studies have shown that levels of BDNF (and BDNF-related PSD-95 signalling) within both the amygdala and hippocampus decrease following fear trials and that BDNF levels reverse towards normal following treatment with ketamine. In such studies, ketamine has been shown to rapidly increase BDNF in the medial PFC and hippocampus, leading to an increase in the number of synapses (synaptic spines) as well as amelioration of the behavioural signs of depression.[7] Fig. 13.2 provides a model of how ketamine, through promoting BDNF (and VEGF), leads to the formation and maturation of synapses.[8]

Ketamine rapidly increases BDNF protein synthesis via a pathway involving AMPA receptor activation and L-type calcium channels, leading to a sustained rise in BDNF through eEF2 kinase signalling and microglial activation, which, through

---

[6] Park and Poo (2013).

[7] Deyama and Kaneda (2023).

[8] Deyama and Duman (2020)).

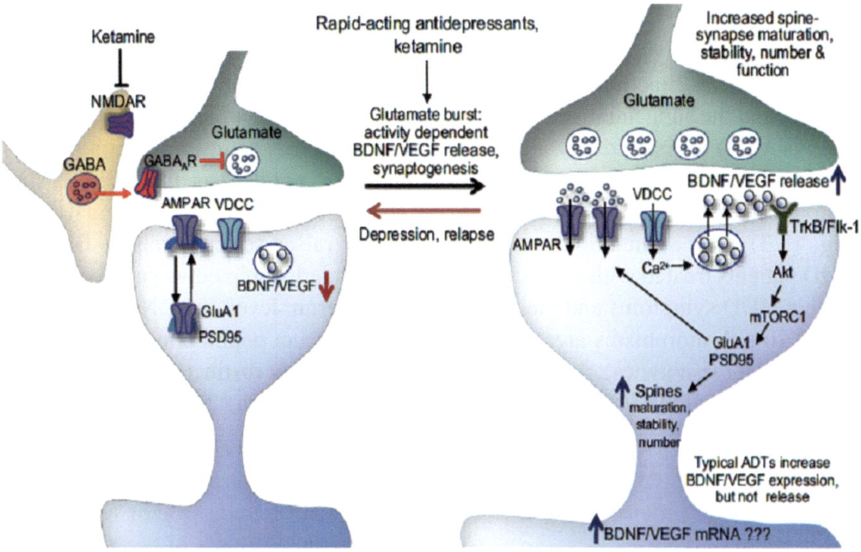

Satoshi Deyamaa,!, Ronald S. Dumanb (2020)

Fig. 13.2 Model for the cellular mechanisms underlying the rapid and sustained antidepressant actions of ketamine. Ketamine blockade of NMDARs on GABAergic interneurons results in disinhibition and a rapid glutamate burst that activates AMPARs. This leads to activation of L-type voltage-dependent Ca2+ channels (VDCCs) and Ca2+ influx that stimulates BDNF and VEGF release. BDNF and VEGF stimulate TrkB and Flk-1, respectively. Activation of TrkB and Flk-1 stimulates the mTORC1 signalling pathway that controls the translation and synthesis of synaptic proteins, including GluA1 and PSD95, that are required for increases in synaptogenesis and spine maturation. These cellular events are associated with the rapid and sustained antidepressant behavioural actions of ketamine. ADT, antidepressant; GABAAR, GABAA receptor. (Reproduced from Deyama and Duman (2020) with permission from Elsevier)

nuclear receptor binding protein 1 (NRBP1) and phosphorylated cAMP response element-binding protein (CREB), give rise to an increased transcription of BDNF.[9] The remarkable rapid antidepressant effects of ketamine are mediated by its rapid release of BDNF, in contrast to the slower BDNF increase induced by serotonin reuptake inhibitors (SSRIs) also used to treat depression.[10] The S-enantiomer of ketamine, esketamine, is now recommended as an intranasal spray, as it has high efficacy in ameliorating MDD.[11]

---

[9] Pardossi et al. (2024).

[10] Duman et al. (2021).

[11] Cui et al. (2024).

13.2.2 Repetitive Transcranial Magnetic Stimulation (rTMS)

There are at least functional connections between BA46 within the dorsolateral PFC (dlPFC) and the subgenual anterior cingulate cortex (ACC), which is strongly implicated in MDD and suicidality. Repetitive transcranial magnetic stimulation (rTMS) involves locating magnetic resonance imaging (MRI) coils over the dlPFC, using anatomical criteria, to best stimulate BA46 for engaging the subgenual ACC (Fig. 13.3). This stimulation using rTMS rapidly raises serum levels of BDNF.[12] MDD patients treated with rTMS show an inverse relationship between the decrease in their MDD symptoms and the increase in their serum level of BDNF.[13]

BDNF polymorphisms are common in the Caucasian population (65% Val/Val, 35% Val/Met genotypes), and these two genotypes show distinct levels of cortical excitability to rTMS.[14] This provides the means of probing the extent to which

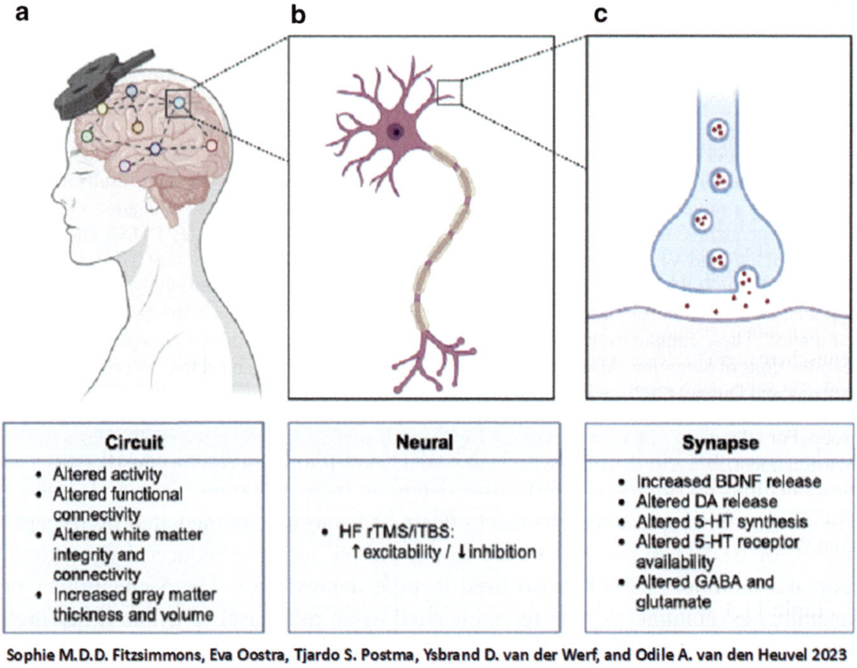

Fig. 13.3 Neuroplasticity effects of repetitive transcranial magnetic stimulation (rTMS) at the level of structural and functional brain circuits (**a**), neurons (**b**) and synapses (**c**). 5-HT, serotonin; BDNF, brain-derived neurotrophic factor; DA, dopamine; GABA, gamma-aminobutyric acid; HF, high-frequency; iTBS, intermittent theta burst stimulation. (Reproduced from Fitzsimmons et al. (2024) with permission from Elsevier)

---

[12] Pan et al. (2023).

[13] Zanardini et al. (2006).

[14] Cheeran et al. (2008).

BDNF is central to the effects of rTMS on ameliorating MDD. rTMS reduces MDD symptoms to a greater extent in patients with Val/Val polymorphism,[15] thereby implicating BDNF in mediating the antidepressant effects of rTMS.

The Hamilton depression rating scale (HAMD) has been used to correlate plasma BDNF levels with depressive severity in patients following rTMS.[16] There is a clear relationship between rTMS-induced improvement in MDD symptoms and serum BDNF levels.[17] There is an increase in serum levels of BDNF in MDD patients after rTMS.[18] MDD patients treated with rTMS have increased levels of serum BDNF compared with sham rTMS.[19] Increases in the duration of rTMS applied to elderly depressed patients increase their serum levels of BDNF.[20] Refractory depression in elderly patients is ameliorated with rTMS that significantly increases BDNF, the increase being negatively correlated with the HAMD-24 score.[21] However, a note of caution is that although measurements of blood serum BDNF in patients with treatment-resistant depression that received rTMS show significant increases in serum BDNF in relation to the decline in their MDD, the overall extent of depression symptoms is not related to the levels of BDNF.[22]

13.2.3 Physical Activity

Studies have found that acute physical exercise increases BDNF in those with and without depression.[23] This increase is efficacious in lowering depression; however, as noted above, there is not a one-to-one relationship between depression severity and BDNF concentration, with BDNF returning to baseline before depression is ameliorated.[24] Pyschotherapies such as prolonged exposure (PE) that lower PTSD symptoms do not affect BDNF levels on their own, but when combined with physical activity, they are more effective in lowering PTSD symptoms and significantly increasing BDNF levels.[25] Clearly, physical activity's effectiveness in increasing BDNF levels contributes significantly to amelioration of both MDD and PTSD.

---

[15] Bocchio-Chiavetto et al. (2008).

[16] Yukimasa et al. (2006).

[17] Zanardini et al. (2006).

[18] Valiuliene et al. (2023).

[19] Pu et al. (2023).

[20] Zhao et al. (2019).

[21] Zhao et al. (2019).

[22] Meshkat et al. (2022).

[23] Szuhany and Otto (2020).

[24] Jemni et al. (2023).

[25] Powers et al. (2015) and Claudino et al. (2020).

13.2.4 Conclusion

Raised BDNF levels are shown to be associated with amelioration of PTSD and MDD symptoms. This suggests that interventions that raise BDNF, such as ketamine, rTMS and physical activity, should be considered in treatment plans for PTSD patients.

13.3 Microglia and Synaptic Plasticity

13.3.1 Early Evidence for Microglial Pruning

The first demonstration that microglia participate directly in synaptic elimination during development came from preclinical models in 2011, showing microglial engulfment of synaptic structures alongside complement deposition.[26] Subsequent live-imaging studies revealed that microglial processes dynamically interact with synapses, responding to neuronal transmitters and remodelling connections in real time.[27]

13.3.2 Location- and Age-Dependent Mechanisms of Pruning

It has recently emerged that one cannot generalize as to the pruning mechanism used by microglia in one part of the brain to another, at least during development. For example, although the complement proteins C1q and C3 were early shown to implement synapse tagging and removal in the early visual thalamus in rodents, it now emerges that fractalkine/CX3CR1 signalling implements tagging and removal in the developing barrel cortex.[28] The development stage also determines the tagging/removal mechanism, as during later stages of development in the visual thalamus, vision-dependent transcription of microglial TNF-like weak inducer of apoptosis (TWEAK) and neuronal Fn14 (the TWEAK receptor) takes over from the C1q and C3 mechanism.[29]

---

[26] Paolicelli et al. (2011) and Faust et al. (2021).

[27] Schafer et al. (2013).

[28] Shen et al. (2024).

[29] Cheadle et al. (2020).

13.3.3 Morphology and Phagocytic Specificity

Synapses are under continual survey by ramified microglia, which use their processes to probe them.[30] Contrary to wholesale synapse engulfment, microglia preferentially internalize small presynaptic fragments (~0.25 μm) via complement-independent pathways, an observation made in hippocampal cultures.[31]

13.3.4 Functional Consequences and Plasticity

Genetic ablation of microglia in preclinical experiments impairs learning and synaptic maturation; similar deficits occur when microglial BDNF is specifically deleted, highlighting BDNF's role in activity-dependent plasticity.[32] However, embryonic microglial depletion produces only minor adult synaptic changes in hippocampal CA1, comparable to adult-onset ablation, suggesting a modulatory rather than obligatory role for microglia in some circuits.[33]

13.3.5 Astrocyte–Microglia Synergy

Astrocytes complement microglial pruning by expressing MERTK and MEGF10 receptors that recognize 'eat-me' signals, and by secreting TGFβ to induce neuronal C1q expression, thereby coordinating the timing and extent of microglial-mediated synapse removal.[34]

13.3.6 Additional Modulatory Pathways

Beyond immune signals, activity-dependent Wnt–Frizzled signalling promotes synapse formation: Wnt7a/Frizzled5 assembles presynaptic terminals, while Wnt7b/Frizzled7 organizes postsynaptic densities.[35] BDNF, released by neurons and microglia, activates TrkB and also potentiates Wnt pathways, further linking trophic signalling to synaptic remodelling.[36]

[30] Miyanishi et al. (2021).
[31] Weinhard et al. (2018).
[32] Parkhurst et al. (2013).
[33] Surala et al. (2024).
[34] Kim and Chung (2023).
[35] Teo and Salinas (2021).
[36] Batool et al. (2019).

13.3.7 Region-Specific Refinement and Disease Implications

Interference with microglial complement pathways (C1q, CX_3CR1, SRPX2) disrupts ocular dominance column formation in the visual cortex, underscoring location-specific pruning requirements.[37] Dysregulation of 'eat-me' signals also contributes to pathological synapse loss—for example, amyloid-β–driven C1q activation in Alzheimer's models leads to excessive pruning, and ischaemic conditions provoke myelin phagocytosis via C3/C3aR pathways.

13.3.8 Conclusion

Microglial phagocytosis of synapses relies on a complex interplay between chemotactic, opsonin-mediated, complement and trophic signals, finely tuned across development, brain regions and physiological states. By balancing elimination and protection, microglia ensure the precise wiring and adaptive plasticity essential for healthy brain function.

13.4 Microglial Molecular Mechanisms in Synapse Loss

In the embryonic and early postnatal brain, microglia adopt an amoeboid morphology, proliferate and migrate throughout the central nervous system (CNS) under the influence of neuronal fractalkine (CX3CL1) and DAP12 signalling (Fig. 13.4). As they mature postnatally, microglia transition to a ramified morphology and upregulate complement receptor 3 (CR3), enabling them to participate actively in synaptic pruning during the postnatal period (Fig. 13.4).[38] In the adult, microglia remain surveillance cells but can release soluble modulators (e.g. BDNF and TNFα) to influence synaptic plasticity.

Microglial phagocytosis of synapses is a highly regulated, multistep process that sculpts neural circuits during development, maintains homeostasis in the adult brain and—when dysregulated—contributes to synaptic loss in disease. The molecular choreography underlying synapse removal involves chemotactic cues ('find-me' signals), surface determinants on neurons ('eat-me' and 'don't-eat-me' signals), soluble and membrane-bound opsonins, specific microglial receptors and downstream engulfment machinery.

---

[37] Cong et al. (2020), Walsh et al. (2020), and Schecter et al. (2017).

[38] Huo et al. (2024).

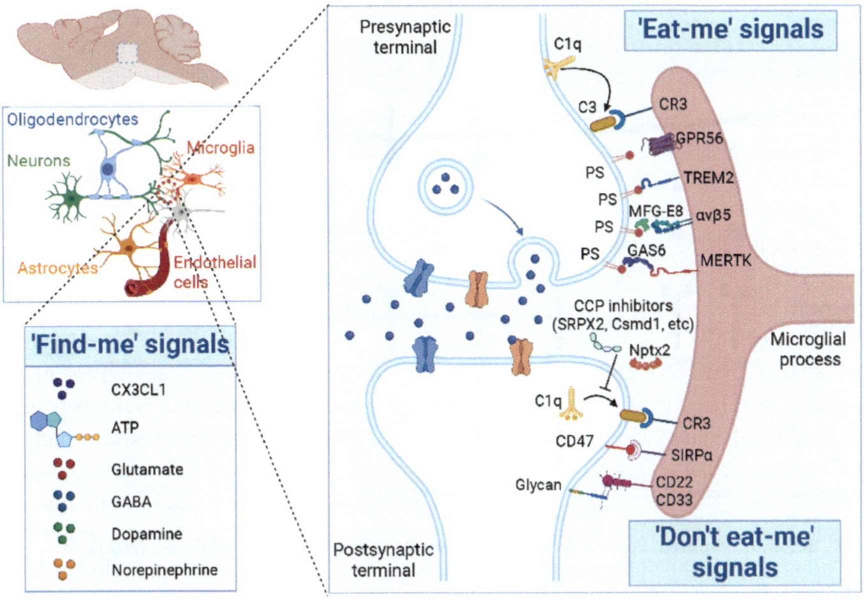

Huo Anran[1]; Wang, Jiali[1]; Li, Qi[1]; Li, Mengqi[1]; Qi, Yuwan[1]; Yin, Qiao[2]; et al., (2024)

Fig. 13.4 Microglial process in synaptic pruning. (i) Neurons secrete distinct find-me signals to recruit microglia towards the immature synapses. (ii) Less active synapses and immature synapses are tagged by eat-me signals phosphatidylserine or complement proteins, which can be recognized by various receptors (CR3, GPR56, TREM2, integrin αvβ5, MERTK) expressed on the microglia. Synaptic pruning occurs through the direct binding-mediated microglial engulfment of tagged synapses. (iii) Microglial phagocytosis is also regulated by don't eat-me signals including classical complement pathway inhibitors (SRPX2, Csmd1, Nptx2), CD47-SIRPα and neuronal polysialylated protein-mediated interaction with CD22 and CD33 to block microglial process. Created with BioRender.com. ATP, adenosine triphosphate; CD22, Siglec-2; CD33, Siglec-3; CD47, cluster of differentiation 47; CR3, complement 3 receptor; CX3CL1, C-X3-C motif chemokine ligand 1; GABA, γ-aminobutyric acid; GPR56, G-protein coupled receptor; MERTK, MER proto-oncogene tyrosine kinase; PS, phosphatidylserine; SIRPα, signal regulatory protein α; SRPX2, sushi repeat-containing protein X-linked 2; TREM2, triggering receptor expressed on myeloid cell 2. (Reproduced from Huo et al. (2024). Creative Commons CC BY 4.0 (https://creativecommons.org/licenses/by/4.0/))

13.4.1 Chemotactic 'Find-Me' Signals

Neurons destined for pruning or in distress release soluble chemoattractants that guide microglial processes towards synaptic sites. Adenosine diphosphate (ADP) binds microglial $P2Y_{12}$ receptors, while CX3CL1 (fractalkine) engages CX3CR1 on microglia, triggering directed migration (Fig. 13.5).[39] This chemotactic response ensures microglia survey active synaptic regions.

[39] Butler et al. (2021), Fig. 1.

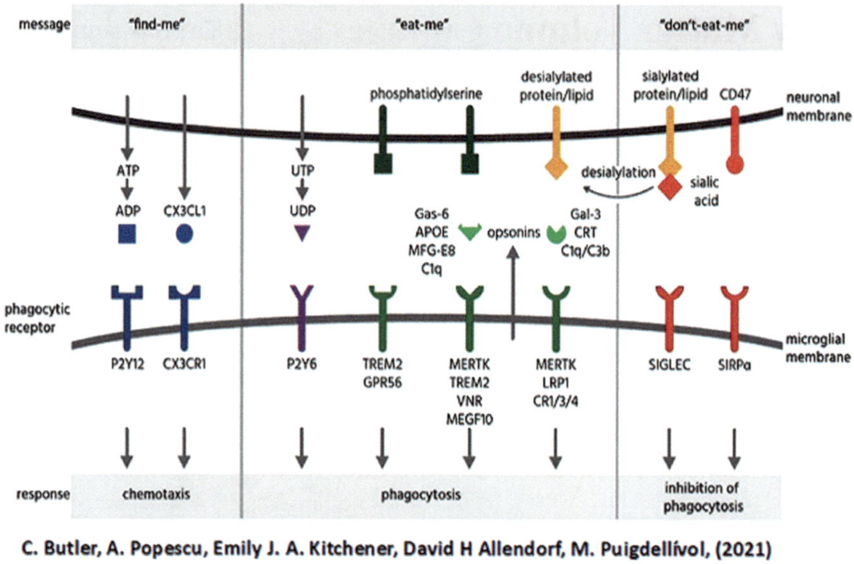

Fig. 13.5 Overview of find-me, eat-me, don't-eat- me signals and opsonins, which regulate microglia phagocytosis of neurons. Find-me signals are chemotactic signals, such as ADP and CX3CL1 (fractalkine) released from neurons and binding to microglial P2Y12 and CX3CR1 receptors, respectively, resulting in chemotaxis of microglia to these neurons. Eat-me signals are released by stressed or dying neurons and induce phagocytosis of neurons expressing them. Such signals include UDP, which activates the P2Y6 receptor on microglia. Phosphatidylserine, when exposed on the surface of neurons, can bind either directly to microglia receptors triggering receptor expressed on myeloid cell 2 (TREM2) and G-protein-coupled receptor 56 (GPR56) or indirectly via binding opsonins, Gas-6apolipoprotein E (APOE), milk fat globule-EGF factor 8 (MFG-E8) and complement component C1q (C1q), which then bind to microglial receptors: receptor tyrosine kinase (MerTK), TREM2, vitronectin (VNR) and multiple EGF-like domain 10 (MEGF10), respectively. Glycosylated proteins and lipids that have been desialylated (had the terminal sialic acid residues removed) can bind opsonins galectin-3(Gal-3), calreticulin (CRT), C1q and complement protein 3b (C3b), which bind to microglial receptors MERTK, low-density lipoprotein receptor-related protein 1 (LRP1) and complement receptors 1/3/4, respectively. Don't-eat-me signals inhibit phagocytosis, and include sialylated glycoproteins and lipids (with terminal sialic acid residue present, recognized by sialic acid binding immunoglobulin-type lectin (Siglecs) receptors) and the protein CD47, which can bind the microglial receptor signal-regulatory protein alpha (SIRPα) to inhibit phagocytosis. Together, these signals regulate microglial phagocytosis of synapses. (Reproduced from Butler et al. (2021) with permission from Wiley)

13.4.2 Eat-Me Versus Don't-Eat-Me Signals

Once in proximity, the balance of opposing surface cues determines whether a synapse is engulfed.

Eat-me signals: Phosphatidylserine (PtdSer) exposure on the outer leaflet of the synaptic membrane can be recognized directly by TREM2 and GPR56 on microglia or indirectly via PtdSer-binding opsonins, such as Gas6, APOE,

MFG-E8 and complement component C1q (all of which bridge to microglial receptors MerTK, TREM2, VNR and MEGF10, respectively) (Fig. 13.5). Soluble nucleotides, like UDP, activate $P2Y_6$ on microglia, further promoting engulfment (Fig. 13.5).

Don't-eat-me signals: Sialylated glycoproteins and lipids engage Siglec receptors on microglia to inhibit phagocytosis. CD47 on synapses binds SIRPα on microglia, delivering a 'self' signal that blocks engulfment (Fig. 13.5).

13.4.3 Complement-Dependent Synaptic Tagging

The classical complement cascade tags synapses for removal in two key steps:

1. C1q binds selective synapses (either directly or via SRPX2 inhibition when absent), initiating formation of the C3 convertase through C4 and C2.
2. The C3 convertase cleaves C3 into C3b, which opsonizes synaptic membranes. Microglial CR3 (CD11b/CD18) binds C3b, triggering engulfment. This pathway is indispensable for developmental pruning and is re-activated pathologically in Alzheimer's disease models, where Aβ drives aberrant C1q deposition (Fig. 13.4).[40]

13.4.4 Fractalkine-Mediated Engulfment

A parallel mechanism involves the chemokine CX3CL1 (fractalkine) (Fig. 13.5). In this mechanism:

- Neuronally expressed CX3CL1 is cleaved by the metalloprotease ADAM10 into a soluble form.
- Soluble CX3CL1 binds CX3CR1 on microglia, not only mediating chemotaxis but also instructing downstream signalling events that promote synaptic engulfment through still-unidentified effectors.
- Disruption of CX3CR1–CX3CL1 signalling leads to deficits in synaptic refinement.

13.4.5 Additional Opsonins and Accessory Pathways

Beyond complement and fractalkine, microglia employ a suite of bridging molecules (Fig. 13.5):

[40] Westacott and Wilkinson (2022).

- Galectin-3 and calreticulin bind desialylated glycoproteins/lipids, linking to microglial MERTK and LRP1.
- Milk fat globule-EGF factor 8 (MFG-E8) binds PtdSer and collaborates with MerTK.
- Multiple EGF-like domain protein 10 (MEGF10) can engage complement opsonins (Fig. 13.5). Astrocyte–microglia cross-talk via TGFβ and IL-33 further refines C1q production by neurons and primes microglia for pruning.

13.4.6 Engulfment and Degradation

The pruning process involves the actin skeleton of microglia promoting envelopment of the labelled synapse by processes of the *glia and internalizing it, or at least part of it, into a phagosome where it is degraded.*

13.4.7 Fn14-Dependent Refinement (Experience-Dependent Phase)

A later refinement phase involves TWEAK binding at specific synapses, signalling their disassembly and removal, complementing earlier complement- and fractalkine-driven pruning.

13.4.8 Conclusion

Three classes of molecular signals coordinate microglial pruning: the 'find-me', 'eat-me' and 'don't-eat-me' (Figs. 13.4 and 13.5). Neurons destined for removal release 'find-me' cues, such as nucleotides (ATP, ADP) and fractalkine, that attract microglial processes. Receptors on these processes then detect phosphatidylserines ('eat-me' signals; see Fig. 13.5) that are externalized on the neuron's synapses, using the receptors TREM2, MERTK, Gas-6, APOE and MFG-E8. Synapses to be removed are also tagged by depositing the complement C1q/C3b that is recognized by CR1/3/4 on the microglia (Fig. 13.5). Large levels of C1q deposition by neurons then lead to inappropriate synapse loss. Signals classified as 'don't-eat-me' and regulate the extent of pruning are sialylated glycoproteins and CD47 on the synapses interacting with SIGLECs and SIRPalpha on the microglia, respectively (Fig. 13.5). Reduced levels of CD47 also lead to inappropriate synapse loss.

References

Bamji SX, Rico B, Kimes N, Reichardt LF (2006) BDNF mobilizes synaptic vesicles and enhances synapse formation by disrupting cadherin–beta-catenin interactions. J Cell Biol 174(2):289–299. https://doi.org/10.1083/jcb.200601087

Batool S, Raza H, Zaidi J, Riaz S, Hasan S, Syed NI (2019) Synapse formation: from cellular and molecular mechanisms to neurodevelopmental and neurodegenerative disorders. J Neurophysiol 121(4):1381–1397. https://doi.org/10.1152/jn.00833.2018

Bocchio-Chiavetto L, Miniussi C, Zanardini R, Gazzoli A, Bignotti S, Specchia C, Gennarelli M (2008) 5-HTTLPR and BDNF Val66Met polymorphisms and response to rTMS treatment in drug-resistant depression. Neurosci Lett 437(2):130–134. https://doi.org/10.1016/j.neulet.2008.04.005

Butler CA, Popescu AS, Kitchener EJA, Allendorf DH, Puigdellívol M, Brown GC (2021) Microglial phagocytosis of neurons in neurodegeneration and its regulation. J Neurochem 158(3):621–639. https://doi.org/10.1111/jnc.15327

Cheadle L et al (2020) Sensory experience engages microglia to shape neural connectivity through a non-phagocytic mechanism. Neuron 108(3):451–468.e9. https://doi.org/10.1016/j.neuron.2020.08.002

Cheeran B et al (2008) A common polymorphism in the brain-derived neurotrophic factor gene (BDNF) modulates human cortical plasticity and the response to rTMS. J Physiol 586(23):5717–5725. https://doi.org/10.1113/jphysiol.2008.159905

Chen CY, Chen YT, Wang JY, Huang YS, Tai CY (2017) Postsynaptic Y654 dephosphorylation of beta-catenin modulates presynaptic vesicle turnover through increased N-cadherin-mediated transsynaptic signaling. Dev Neurobiol 77(1):61–74. https://doi.org/10.1002/dneu.22411

Claudino FCA, Gonçalves L, Schuch FB, Martins HRS, da Rocha NS (2020) The effects of individual psychotherapy in BDNF levels of patients with mental disorders: a systematic review. Front Psych 11:445. https://doi.org/10.3389/fpsyt.2020.00445

Cong Q, Soteros BM, Wollet M, Kim JH, Sia GM (2020) The endogenous neuronal complement inhibitor SRPX2 protects against complement-mediated synapse elimination during development. Nat Neurosci 23(9):1067–1078. https://doi.org/10.1038/s41593-020-0672-0

Cui L et al (2024) Major depressive disorder: hypothesis, mechanism, prevention and treatment. Signal Transduct Target Ther 9(1):30. https://doi.org/10.1038/s41392-024-01738-y

Deyama S, Duman RS (2020) Neurotrophic mechanisms underlying the rapid and sustained antidepressant actions of ketamine. Pharmacol Biochem Behav 188:172837. https://doi.org/10.1016/j.pbb.2019.172837

Deyama S, Kaneda K (2023) Role of neurotrophic and growth factors in the rapid and sustained antidepressant actions of ketamine. Neuropharmacology 224:109335. https://doi.org/10.1016/j.neuropharm.2022.109335

Duman RS, Deyama S, Fogaça MV (2021) Role of BDNF in the pathophysiology and treatment of depression: activity-dependent effects distinguish rapid-acting antidepressants. Eur J Neurosci 53(1):126–139. https://doi.org/10.1111/ejn.14630

Faust TE, Gunner G, Schafer DP (2021) Mechanisms governing activity-dependent synaptic pruning in the developing mammalian CNS. Nat Rev Neurosci 22(11):657–673. https://doi.org/10.1038/s41583-021-00507-y

Fitzsimmons SMD et al (2024) Repetitive transcranial magnetic stimulation-induced neuroplasticity and the treatment of psychiatric disorders: state of the evidence and future opportunities. Biol Psychiatry 95(6):592–600. https://doi.org/10.1016/j.biopsych.2023.11.016

Huo A, Wang J, Li Q, Li M, Qi Y, Yin Q, Luo W, Shi J, Cong Q (2024) Molecular mechanisms underlying microglial sensing and phagocytosis in synaptic pruning. Neural Regen Res 19(6):1284–1290. https://doi.org/10.4103/1673-5374.385854

Jemni M et al (2023) Exercise improves depression through positive modulation of brain-derived neurotrophic factor (BDNF): a review based on 100 manuscripts over 20 years. Front Physiol 14:1102526. https://doi.org/10.3389/fphys.2023.1102526

Kim NS, Chung WS (2023) Astrocytes regulate neuronal network activity by mediating synapse remodeling. Neurosci Res 187:3–13. https://doi.org/10.1016/j.neures.2022.09.007

Meshkat S et al (2022) Brain-derived neurotrophic factor (BDNF) as a biomarker of treatment response in patients with treatment-resistant depression: a systematic review and meta-analysis. Psychiatry Res 317:114857. https://doi.org/10.1016/j.psychres.2022.114857

Miyanishi K, Sato A, Kihara N, Utsunomiya R, Tanaka J (2021) Synaptic elimination by microglia and disturbed higher brain functions. Neurochem Int 142:104901. https://doi.org/10.1016/j.neuint.2020.104901

Pan F et al (2023) Effects of neuronavigation-guided rTMS on serum BDNF, TrkB and VGF levels in depressive patients with suicidal ideation. J Affect Disord 323:617–623. https://doi.org/10.1016/j.jad.2022.11.059

Paolicelli RC et al (2011) Synaptic pruning by microglia is necessary for normal brain development. Science 333(6048):1456–1458. https://doi.org/10.1126/science.1202529

Pardossi S, Fagiolini A, Cuomo A (2024) Variations in BDNF and their role in the neurotrophic antidepressant mechanisms of ketamine and esketamine: a review. Int J Mol Sci 25(23):13098. https://doi.org/10.3390/ijms252313098

Park H, Poo MM (2013) Neurotrophin regulation of neural circuit development and function. Nat Rev Neurosci 14(1):7–23. https://doi.org/10.1038/nrn3379

Parkhurst CN et al (2013) Microglia promote learning-dependent synapse formation through brain-derived neurotrophic factor. Cell 155(7):1596–1609. https://doi.org/10.1016/j.cell.2013.11.030

Powers A, Cross D, Fani N, Bradley B (2015) PTSD, emotion dysregulation, and dissociative symptoms in a highly traumatized sample. J Psychiatr Res 61:174–179. https://doi.org/10.1016/j.jpsychires.2014.12.011

Pu Z, Hou Q, Yan H, Lin Y, Guo Z (2023) Efficacy of repetitive transcranial magnetic stimulation and agomelatine on sleep quality and biomarkers of adult patients with mild to moderate depressive disorder. J Affect Disord 323:55–61. https://doi.org/10.1016/j.jad.2022.11.062

Schafer DP, Lehrman EK, Stevens B (2013) The "quad-partite" synapse: microglia–synapse interactions in the developing and mature CNS. Glia 61(1):24–36. https://doi.org/10.1002/glia.22389

Schecter RW et al (2017) Experience-dependent synaptic plasticity in V1 occurs without microglial CX3CR1. J Neurosci 37(44):10541–10553. https://doi.org/10.1523/JNEUROSCI.2679-16.2017

Shen FS et al (2024) Emerging evidence of context-dependent synapse elimination by phagocytes in the CNS. J Leukoc Biol 116(3):511–522. https://doi.org/10.1093/jleuko/qiae098

Surala M et al (2024) Lifelong absence of microglia alters hippocampal glutamatergic networks but not synapse and spine density. EMBO Rep 25(5):2348–2374. https://doi.org/10.1038/s44319-024-00130-9

Szuhany KL, Otto MW (2020) Assessing BDNF as a mediator of the effects of exercise on depression. J Psychiatr Res 123:114–118. https://doi.org/10.1016/j.jpsychires.2020.02.003

Teo S, Salinas PC (2021) Wnt–Frizzled signaling regulates activity-mediated synapse formation. Front Mol Neurosci 14:683035. https://doi.org/10.3389/fnmol.2021.683035

Valiuliene G et al (2023) Anti-neuroinflammatory microRNA-146a-5p as a potential biomarker for neuronavigation-guided rTMS therapy success in medication-resistant depression disorder. Biomed Pharmacother 166:115313. https://doi.org/10.1016/j.biopha.2023.115313

Walsh EI, Smith L, Northey J, Rattray B, Cherbuin N (2020) Towards an understanding of the physical activity–BDNF–cognition triumvirate: a review of associations and dosage. Ageing Res Rev 60:101044. https://doi.org/10.1016/j.arr.2020.101044

Weinhard L et al (2018) Microglia remodel synapses by presynaptic trogocytosis and spine-head filopodia induction. Nat Commun 9(1):1228. https://doi.org/10.1038/s41467-018-03566-5

Westacott LJ, Wilkinson LS (2022) Complement dependent synaptic reorganisation during critical periods of brain development and risk for psychiatric disorder. Front Neurosci 16:840266. https://doi.org/10.3389/fnins.2022.840266

Yukimasa T et al (2006) High-frequency repetitive transcranial magnetic stimulation improves refractory depression by influencing catecholamine and brain-derived neurotrophic factors. Pharmacopsychiatry 39(2):52–59. https://doi.org/10.1055/s-2006-931542

Zanardini R et al (2006) Effect of repetitive transcranial magnetic stimulation on serum brain-derived neurotrophic factor in drug-resistant depressed patients. J Affect Disord 91(1):83–86. https://doi.org/10.1016/j.jad.2005.12.029

Zhao X, Li Y, Tian Q, Zhu B, Zhao Z (2019) Repetitive transcranial magnetic stimulation increases serum brain-derived neurotrophic factor and decreases interleukin-1β and tumor necrosis factor-α in elderly patients with refractory depression. J Int Med Res 47(5):1848–1855. https://doi.org/10.1177/0300060518817417

Part V
Ameliorating Veteran's PTSD with Neurotherapies

Chapter 14
Abnormalities in Resting State Networks in PTSD and Treatment with Transcranial Magnetic Stimulation (TMS)

14.1 Abnormalities of Functional Activity and Connectivity of Resting State Networks in PTSD

14.1.1 The Default Mode Network (DMN), the Central Executive Network (CEN) and the Salience Network (SN)

At the resting state network (RNS) level of the brain, PTSD leads to the default mode network (DMN) and the central executive network (CEN) being functionally hypoactive with weak functional connectivity, while the salience network (SN), in contrast, is hyperactive and hyperconnected (Sect. 10.3). This suggests that the SN has a 'low threshold for perceived saliency', leading to inappropriate modulation of DMN CEN interaction (Fig. 14.1).[1]

A variety of meta-analyses agree with this summary of the extent to which PTSD modifies RSNs (summarized in Table 14.1). There is decreased DMN connectivity and enhanced SN connectivity with hyperfunctional activity in the anterior cingulate cortex (ACC) and the amygdala/parahippocampus.[2] There is hypoactivity in the ventromedial prefrontal cortex (vmPFC), which is a core node of the DMN, and hyperactivity in the mid–/dorsal ACC, a core node of the SN, with the amygdala also hyperactive.[3] Meta-analysis agrees with this, indicating a hypoactive medial prefrontal cortex (mPFC), which is a core node of the DMN, accompanied by a hyperactive amygdala in PTSD[4]; and that PTSD hypoactivity occurs in the mPFC

[1] Akiki et al. (2017).

[2] Koch et al. (2016).

[3] Hayes et al. (2012) and Pitman et al. (2012).

[4] Patel et al. (2012).

M. Bennett, *Veteran's Stress, Trauma and Synapse Rehabilitation*, https://doi.org/10.1007/978-981-95-9594-5_14

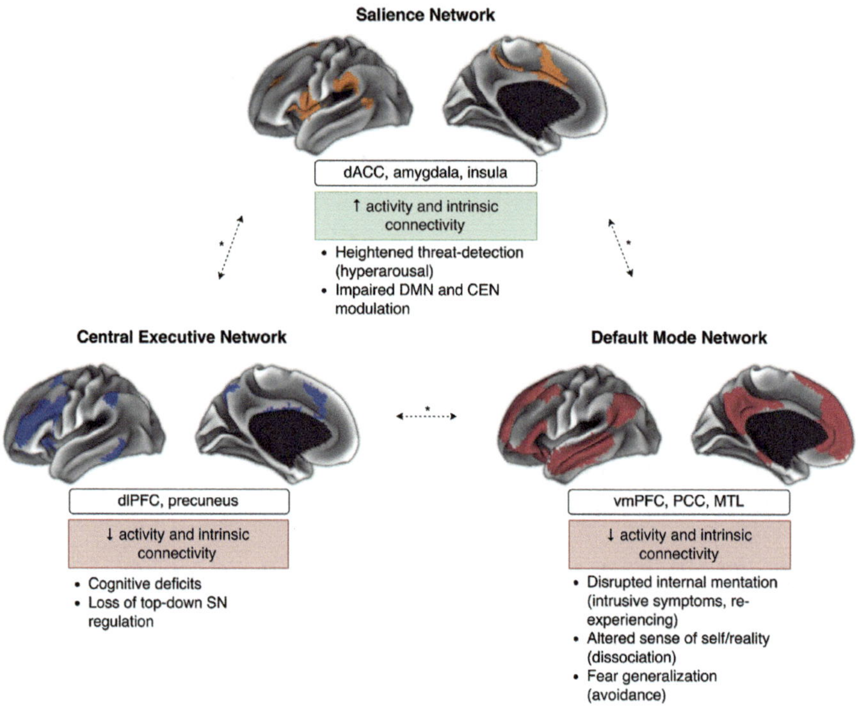

Fig. 14.1 Model of PTSD. The figure shows the cortical representations of the salience network (SN; orange), default mode network (DMN; red) and central executive network (CEN; blue) in healthy individuals, notable region of interests (ROIs) within these networks, the changes in PTSD and the putative resulting phenotypic abnormalities. Alterations within and between the SN, the CEN and the DMN may underlie the psychopathology of PTSD. The SN is hyperconnected and hyperactive and has a low threshold for perceived saliency (underlying symptoms of hyperarousal) and is incapable of efficient DMN–CEN modulation (i.e. switching between task-relevant and task-irrelevant behavior); the CEN is weakly interconnected and hypoactive (paralleling impaired cognition) and is incapable of top-down SN regulation; finally, the DMN is weakly interconnected and hypoactive, resulting in disrupted ability to maintain a calm inner state (intrusive symptoms), altered sense of self/world (dissociation) and fear generalization (avoidance; mediated by the hippocampus). The asterisk denotes an altered between-network connectivity. PTSD may be characterized by a weakly connected and hypoactive default mode network (DMN) and central executive network (CEN) that are putatively destabilized by an overactive and hyperconnected salience network (SN), which appears to have a low threshold for perceived saliency, and in efficient DMN–CEN modulation. (Reproduced from Akiki et al. (2017) with permission from Springer Nature)

and hyperactivity in the posterior insula (a core node of the SN).[5] Another study observed reduced functional connectivity within the DMN (e.g. between the vmPFC and the rostral ACC) but enhanced connectivity within the SN (e.g. between the insula and other areas of the SN), as well as between the DMN and SN, and increased

[5]Wang et al. (2016).

Table 14.1 Alterations in functional activity and connectivity of resting state network in PTSD patients

Network role	Default mode network (DMN)	Salience network (SN)	Central executive network (CEN)
Functional activity	− vmPFC[a–d] − PCC[a–d] − hippocampus[b] + H[c] + anterior hippocampus[c]	+ amygdala[a–c] + dACC[c] + anterior insula[a,c,d]	− dlPFC[a,c] − precuneus[c]
Intrinsic functional connectivity (node–node)	− vmPFC–hippocampus[e] − vmPFC–posterior hippocampus[f] − PCC–posterior hippocampus[f] − PCC–hippocampus[g,h] − PCC–vmPFC[g,i]	Amygdala–insula[e,j] Amygdala–dACC[k] − dACC–SN[i]	− precuneus–CEN[e,i] − premotor cortex–dlPFC[l]
Extrinsic functional connectivity (node–node)	+ insula–hippocampus[e] + amygdala–vmPFC[k]	+ insula–hippocampus[e] + amygdala–vmPFC[k] + SN–CEN[m]	+ SN–CEN[m]

+, increase; −, decrease; dACC, dorsal anterior cingulate cortex; dlPFC, dorsolateral prefrontal cortex; PCC, posterior cingulate cortex; vmPFC, ventromedial prefrontal cortex
Notes:
[a]Koch et al. (2016)
[b]Hayes et al. (2012)
[c]Patel et al. (2012)
[d]Wang et al. (2016)
[e]Sripada et al. (2012)
[f]Chen and Etkin (2013)
[g]Bluhm et al. (2009)
[h]Viard et al. (2019)
[i]St Jacques et al. (2013)
[j]Rabinak et al. (2011)
[k]Brown et al. (2014)
[l]Cisler et al. (2013)
[m]Rabellino et al. (2015)

connectivity in the amygdala.[6] Each of these RSNs is likely to be associated 'with specific clinical symptoms observed in PTSD, including cognitive dysfunction (CEN), increased and decreased arousal/interoception (SN) and an altered sense of self (DMN)'.[7] In summary:

---

[6] Sripada et al. (2012).

[7] Lanius et al. (2015).

> The most commonly cited neuroanatomical model of PTSD focuses on altered threat detection and fear learning systems and suggests that the disorder is a result of a hyper-responsive amygdala, insula, and dorsal anterior cingulate cortex to threat stimuli, inadequate inhibition of the amygdala by the ventromedial prefrontal cortex, and an inability, or resistance to, extinction and its recall.[8]

Table 14.2 summarizes the data in Table 14.1 to make explicit that DMN hypo-connectivity impairs self-referential processing, SN hyperactivity drives heightened emotional reactivity and CEN dysfunction contributes to cognitive deficits.

14.1.2 Abnormalities of the DMN Associated with PTSD

A large number of studies using functional magnetic resonance imaging (fMRI) have shown that there is reduced functional activity and intrinsic functional connectivity in DMN in patients with PTSD, leading to an altered sense of self.[9] When this involves the mPFC component of the DMN,[10] it leads to loss of functional control of the amygdala, with consequences for emotional balance.[11] Decreasing functional connectivity strength in the DMN is correlated with an increase in PTSD as

Table 14.2 Effect of abnormalities in resting state networks associated with PTSD symptoms

Network role	Default mode network (DMN)	Salience network (SN)	Central executive network (CEN)
Functional activity	Decreased during resting state	Increased, reflecting hypervigilance and threat detection	Decreased activity, linked to impaired cognitive control
Intrinsic functional connectivity	Decreased, particularly in the medial prefrontal cortex and hippocampus	Increased, especially between the amygdala and insula	Decreased intrinsic connectivity within dorsolateral prefrontal cortex
Extrinsic functional connectivity	Weakened connections with other networks, including SN and CEN	Stronger coupling with DMN (e.g. between the amygdala and the DMN) and CEN under stress	Increased coupling with SN, possibly to compensate for trauma-related hyperactivity
Associated PTSD symptoms	Altered sense of self	Increased or decreased arousal or interoception	Cognitive dysfunction

Sources: Table 1; Lanius et al. (2015)

[8] Shin and Liberzon (2010); Graham and Milad (2011); Vuper et al. (2021); Hajong (2021)

[9] Sripada et al. (2012), Zhang et al. (2015), Zandvakili et al. (2020), Bluhm et al. (2009), Akiki et al. (2017).

[10] Sripada et al. (2012), Akiki et al. (2017), and Zhang et al. (2015); but see Viard et al. (2019).

[11] Hajong (2021).

measured using the clinician-administered PTSD scale (CAPS) (Fig. 14.2), with this decrease in strength primarily due to decreased functional connectivity in the mPFC.[12]

Given the central role of the DMN in relation to PTSD, Table 14.3 spells out the normal functions of elements of the DMN, and Table 14.4 sets out what goes awry in the connectivity of these elements leading to the clinical signs in PTSD.

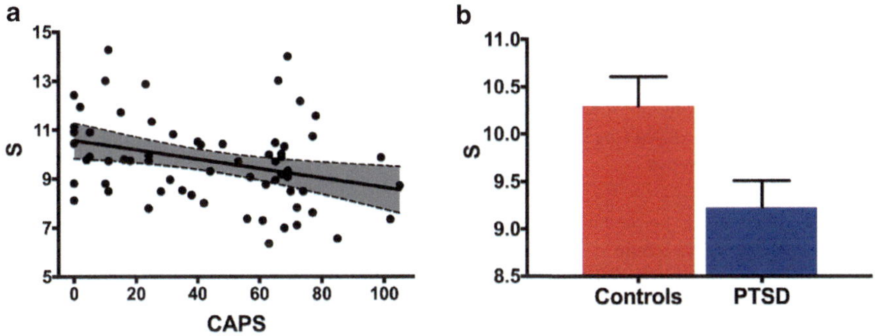

Fig. 14.2 Correlation between posttraumatic stress disorder (PTSD) and default mode network (DMN) connectivity strength. (**a**) Scatter plot depicting the correlation between PTSD severity, as measured by the clinician-administered PTSD scale (CAPS), and DMN overall connectivity strength (S). (**b**) Bar graph depicting average DMN overall connectivity strength (S) compared between control and PTSD patients. (Reproduced from Akiki et al. (2017) with permission from Elsevier)

Table 14.3 Regions of the default mode network (DMN) and associated functions

Region of the DMN	Associated functions
Anterior cingulate cortex	Detect and correct errors, regulating both cognitive and emotional processing
Medial prefrontal cortex	Information processing relevant to self and considering the minds of other people
Parietal lobes	Somatosensory perception, bodily perception, visual–spatial orientation, memory, symbolic synthesis and cross-modal matching
Posterior cingulate cortex	Evaluative functions and mediation of interactions of emotional and memory-related processes
Precuneus	Mental imagery strategies related to the self; facilitation of successful episodic memory retrieval
Temporal lobes	Role in memory for past events, affecting both autobiographical (i.e. episodic and semantic) and non-autobiographical (i.e. public events and general semantic knowledge) memory
Temporo-parietal junction	Theory of mind; empathy

Source: Chan (2016), Table 1. Reproduced from https://digitalcommons.pepperdine.edu/etd/660

[12] Akiki et al. (2017)

Table 14.4 Neuropsychological profile of neural dysfunction in the default mode network (traumatized resting state)

Neural dysfunction	Description of effect
Reduced connectivity of the medial prefrontal cortex (mPFC) and limbic regions. Hypoactivity in the mPFC and hyperactivity in the amygdala	Increased emotional reactivity; aberrant self-referential processing
Increased connectivity and activation of the posterior cingulate cortex (PCC) with the right amygdala	Threat, survival and negative-biased processing in evaluation of environment and emotion-mediated memory
Decreased connectivity between the PCC and precuneus, bilaterial parietal cortex and mPFC	Inaccurate (biased) and decreased capacity for evaluation of self, prospection, social cognition and moral sensitivity
Reduced left hippocampal functioning. Increase in the right hippocampus	Reduced autobiographical recall (context-dependent); reduced memory for locations (priming for action)
Decreased activity in the precuneus, mPFC and right superior parietal lobule	Reduced capacity for self-reflection; biased prospection
Reduced connectivity between the anterior cingulate cortex and mPFC and medial temporal gyrus	Dissonance between 'oops Centre' and self-function, social cognition and moral sensitivity
Increased insula activity and connectivity with the amygdala (salience network)	Poor interoception—Exaggerated bodily response
Reduced connectivity between the insula and PCC	Dissonance between interoception and evaluative functions and emotion-mediated memory
Dissociated left and right amygdala	Left: Pleasant/unpleasant emotions; reward system Right: Negative emotions, fear and sadness

Source: Chan (2016), Table 6. Reproduced from https://digitalcommons.pepperdine.edu/etd/660

14.2 Transcranial Magnetic Stimulation (TMS) Used to Normalize RSN Functional Activity and Connectivity

14.2.1 TMS Used to Treat Psychiatric Disease

A large number of studies show that transcranial magnetic stimulation (TMS) can increase the functional connectivity between the three principal RSNs of interest in PTSD. It was first ascertained that in major depressive disorder (MDD), unlike in PTSD, the functional connectivity is elevated within the DMN, and this can be lowered to the normal level by TMS stimulation of the dlPFC.[13] Subsequently, it was shown in patients with comorbid MDD and PTSD that clinical improvement could be elicited with TMS of appropriate sites in the dlPFC that reduced functional connectivity between the DMN and the subgenual ACC (sgACC), the insula (SN) and

[13] Liston et al. (2014).

the dlPFC (CEN) and reduced connectivity between the SN and the hippocampus.[14]

TMS involves positioning of the stimulating coil on the brain. The sgACC (BA25) has long been known to play a central role in the development of MDD, but this deep region of the cortex is inaccessible to direct stimulation using TMS as it is 4–5 cm below the scalp, and standard TMS coils reach to less than 3 cm. However, BA25 can be reached indirectly, for a well-established feedforward projection exists from the dlPFC (BA10) to the pregenual ACC (BA32) and from there to the sgACC. Thus, recording the activity in BA25 with fMRI while positioning the TMS coil at different sites within the dlPFC area until optimal changes in the activity of BA25 have been obtained has been shown to decrease MDD in patients.[15] Given the overlap between MDD symptoms and those of PTSD (see Sect. 2.4), and that there is about 50% comorbidity, it is very likely that positioning the TMS coil over the lateral prefrontal cortex (PFC), thereby modulating BA25 activity, will not only help alleviate symptoms of MDD but also, to some extent, those of PTSD. Early studies show this anticipated decrease in PTSD symptoms to be the case, as have more recent studies.

Most recently, the TMS coil has been positioned over the lateral PFC according to maximum modulation of BA25 subgenual activity, determined using fMRI, and surprisingly obtained significant decreases in the hyperarousal symptoms of PTSD.[16] However, a powerful TMS approach to alleviating PTSD is undoubtedly to modulate activity in the amygdala, which receives a direct projection from the lateral PFC. Positioning the TMS electrode on lateral PFC such that it produces maximum modulation of the amygdala according to fMRI criteria is now possible.

14.2.2 Stimulating the 'PTSD Network': The Amygdala, Medial Prefrontal Cortex, and Anterolateral Temporal Lobe

A very recent significant study has shown that a 'PTSD network' exists consisting of the amygdala, the mPFC and the anterolateral temporal lobe, with particular emphasis on the mPFC.[17] This study found that certain lesions to the heads of traumatized victims protect them from developing PTSD.[18] Key among these are the mPFC and amygdala together with the anterolateral temporal lobe—that is, the PTSD network. When these areas are stimulated with TMS, particularly the mPFC, PTSD symptoms are dramatically alleviated. Because the mPFC and several nuclei

---

[14] Philip et al. (2019).

[15] Cole et al. (2022).

[16] Cash et al. 2019.

[17] Siddiqi et al. (2024).

[18] Ibid.

of the amygdala are principal components of the DMN, this is consistent with the abnormalities in DMN functional activity in PTSD (see Sect. 14.1). Nuclei of the amygdala are also part of the SN, which is consistent with the SN abnormalities. The discovery of this PTSD network suggests that TMS remedial treatment for PTSD should concentrate on the amygdala and mPFC. Like the sgACC, the amygdala's depth at 4–5 cm beneath the scalp means it cannot be reached directly, but it too can be reached by appropriate positioning of the TMS electrode over the dlPFC. The mPFC can be reached directly as it is 2–3 cm beneath the scalp, although the vmPFC might require coils with a deep reach.[19] The anterolateral temporal lobe component of the PTSD network can be reached directly with TMS stimulation.[20]

14.2.3 Stimulation of the Dorsolateral Prefrontal Cortex to Treat PTSD

Because the dlPFC projects to the pregenual ACC (BA32), which in turn projects to the sgACC (BA25), TMS stimulation of the dlPFC can reach the sgACC.[21] In an early use of TMS to alleviate PTSD symptoms, the TMS coils were placed on the lateral PFC, and a stimulation protocol was used five times per week for 2 weeks, resulting in a 50% decrease in PTSD symptoms.[22] Subsequent focus on the dlPFC showed similar or better results. One study showed improvement of about 60% in PTSD symptoms, lasting more than 3 months, following TMS of the dlPFC.[23] Other studies show that TMS of the right dlPFC gives a 50–60% reduction in PTSD symptoms.[24] A meta-analysis of all results before 2019 showed that stimulation of the dlPFC with TMS reduces PTSD symptoms.[25] More recently, TMS stimulation of the right dlPFC was found to improve PTSD symptoms with a standard mean difference of 0.72 (a medium difference).[26] Theta burst stimulation (frequencies of about 50 Hz repeated every 200 ms) of the dlPFC with TMS produced a 50% decrease in symptoms on the Davidson trauma scale.[27]

The use of fMRI to determine functional connectivity between sites in the dlPFC and the sgACC, up to 2019, was the subject of a review to determine

[19] But see Kearney-Ramos et al. (2018).

[20] See, e.g. Woollams et al. (2017).

[21] Joyce et al. (2020).

[22] Taghva et al. (2015); see also Makale et al. (2023).

[23] Boggio et al. (2010).

[24] Berlim and Van Den Eynde (2014) and Ahmadizadeh and Rezaei (2018).

[25] Koek et al. (2019).

[26] McGirr et al. (2021).

[27] Chang et al. (2023).

the optimal choice of sites for TMS.[28] Stimulation with TMS of the lateral PFC using fMRI guidance to find stimulation sites that have maximum connectivity to the sgACC was found to provide a 70% decrease in PTSD symptoms, specifically in hyperarousal.[29] There is a direct nerve fiber pathway from the vlPFC to the amygdala, such that sites on the vlPFC can be found for TMS stimulation of trauma-located sites. Recent research suggests that the electroencephalography (EEG) beta power (at 13–30 Hz) is also increased with TMS stimulation of the left dlPFC and that this is negatively correlated with both PTSD and MDD symptoms, suggesting that beta power might at least act as a biomarker of PTSD.[30]

14.3 Other Therapies for Normalizing Abnormalities in RSNs

14.3.1 Trauma-Focused Cognitive Behaviour Therapy (TF-CBT)

Trauma-focused cognitive behaviour therapy (TF-CBT) is effective in enhancing functional connectivity both within and between the DMN, CEN and SN,[31] with the clinical result that both therapies show major improvements in patients over waitlist/usual care.[32] The lower the pretreatment functional connectivity in the DMN and the CEN, the higher the response level to TF-CBT, interpreted as improvements in functional connectivity within and between these RSNs, which are concerned with self-awareness, cognitive control and external vigilance.[33]

14.3.2 Mindfulness-Based Interventions

Mindfulness-based therapies have been shown to increase functional connectivity between the DMN, the CEN and the SN to normal.[34] This correlates to mindfulness ameliorating PTSD.

[28] McDonald and van Rooij (2019).

[29] Oathes et al. (2024).

[30] Morriss et al. (2024).

[31] Charquero-Ballester et al. (2022).

[32] Bisson et al. (2013).

[33] Korgaonkar et al. (2020).

[34] King et al. (2016) and Boyd et al. (2018).

14.3.3 Neurofeedback

Recent meta-analyses of new studies of fMRI-guided neurofeedback (NF) (see Chap. 15) of deep brain areas like the amygdala have significantly enhanced the standing of this kind of therapy.[35]

Activity of the amygdala (associated with the DMN) is an indicator of the extent of a patient's PTSD, so that fMRI determination of this activity provides a measure of their PTSD.[36] Using fMRI-determined amygdala activity in a neurofeedback paradigm to decrease its functional activity leads also to increased activity of the dlPFC and hence of the CEN and normalization of the DMN activity.[37] This is consistent with the observation that decreasing posterior cingulate cortex (PCC) activity as measured with fMRI following EEG-NF leads to an increase in the connectivity between the amygdala and the PCC.[38] Interestingly, when the activity of the sgACC, which is well correlated with MDD symptoms as well as the hyperarousal and avoidance symptoms of PTSD, is subjected to fMRI-guided TMS; these PTSD symptoms are significantly ameliorated.[39] Enhancing low-resolution electromagnetic tomography to a higher resolution so as to improve the site-directed NF provides a significant normalization of DMN, CEN and SN activity and connectivity and improvement in the clinical symptoms of responding PTSD patients.[40] Using alpha desynchronizing neurofeedback provides greater functional connectivity in the SN and DMN (mPFC).[41]

14.3.4 Pharmacological Interventions

The pharmacological intervention that has become of principal interest recently is ketamine. This drug increases the functional connectivity of the vmPFC (DMN) with the amygdala, as does oxytocin,[42] decreasing its activity and thus ameliorating PTSD symptoms.[43] Ketamine has also been found to decrease functional connectivity between the hippocampus (DMN) and the amygdala, thereby ameliorating PTSD symptoms in patients.[44] For those with MDD, ketamine normalizes the connectivity between the DMN and the insula (SN), leading to a decrease in anxiety.[45]

[35] Voigt et al. (2024).

[36] Zhao et al. (2023).

[37] Nicholson et al. (2018).

[38] Lieberman et al. (2023).

[39] Oathes et al. (2024).

[40] Bell et al. (2019), Nicholson et al. (2016), and Shaw et al. (2023). For a meta-analysis, see Askovic et al. (2023).

[41] Kluetsch et al. (2014).

[42] Lotfinia et al. (2023).

[43] Norbury et al. (2021).

[44] Duek et al. (2023).

[45] Evans et al. (2018) and Vasavada et al. (2021).

14.4 Comparison of Therapies

The effects of different therapies on changes in functional connectivity within and between different RSNs are shown in detail in Table 14.5 and simplified and summarized in Table 14.6.

In general, a number of therapeutic methods—TF-CBT, mindfulness, NF, ketamine and TMS—ameliorate PTSD by addressing the network imbalances that characterize PTSD, described in Sect. 14.1. Specifically, they enhance the weak connectivity and hypoactivity of the DMN and CEN while reducing the hyperconnectivity and overactivity of the SN, thereby stabilizing the low-threshold salience detection and restoring efficient DMN–CEN modulation.

Table 14.5 Effects of different therapies on changes in functional connectivity within and between different resting state networks

TF-CBT	Mindfulness (MBET)	Neurofeedback (NFB)	Pharmacology (P)	Transcranial magnetic stimulation (TMS)
DMN[1] (iFc), DMN-SN[1](eFc), DMN-CEN[1](eFc), SN[1] (iFc), SN-CEN[1] (eFc), CEN[1] (iFc) *DMN[2] (eFc)	DMN[5] (F) (vmPFC), DMN-SN[4] (eFc), DMN-CEN[4] (eFc), DMN[5](F) (rACC), #DMN[8] (iFc), #DMN[9] (iFc) (PCC-vmPFC), #DMN-SN[9] (eFc)(PCC-A)	DMN[10] (F), DMN[7] (iFc), DMN[6] (F) (PCC, mPFC), DMN[7] (iFc), SN[6](-F)(aI), SN[10] (-F), CEN[10] (F), CEN[14]((F) (dlPFC), LN[11]((iFc)(A), LN[20] (-F) (A) Meta-analysis/review[24]	DMN[15] (F) (vmPFC)[THC], DMN-SN[15] (eFc) (vmPFC-A)[O], DMN-SN[16](eFc) (vmPFC-A)[K], SN[15](F)(A)[O], SN[15](-F)(A) [THC], CEN[15](F)(dlPFC) [O], LN[17] (iFc)(A-H) [K]	DMN-sgACC[18] (-eFc), SN-sgACC[18] (I-sgACC) (-eFc), SN-DMN[18](dACC-pH) (-eFc) CEN-sgACC[18] (dlPFC-sgACC)(-eFc), CEN-LN[21] (dlPFC-A) (-eFc) CEN-LN[22] (dlPFC-A) (-eFc) CEN-LN[23](vlPFC-A) (-eFc)

F functional activity, *eFc* extrinsic functional connectivity, *iFc* intrinsic functional activity, *THC* delta-9-tetrahydrocannabinol, *O* oxytocin, *K* ketamine, * cognitive processing therapy, # other therapies, *LN* limbic network

Examples of how to read this table are as follows: "DMN[1] (iFc)" refers to an increase in functional connectivity (iFc) within the DMN, the superscript giving the numbered reference below; "DMN-SN[1](eFc)" refers to an increase in extrinsic functional connectivity (eFc) between DMN and SN, the superscript giving the numbered reference below; "DMN[5] (F)(vmPFC)" refers to an increase in the functional activity (F) of the vmPFC component of the DMN, given in reference 5. Square brackets, where they appear in the Pharmacology (P) column, indicate the pharmacological treatment type. Negative values of eFc indicate a decrease in extrinsic functional connectivity; negative values of F indicate a decrease in functional activity

References for Table 14.5

1. Korgaonkar et al. (2020)
2. Vuper et al. (2021) [cognitive processing therapy]
3. Vuper (2020)
4. Boyd et al. (2018).
5. King et al. (2016) [Improvement in avoidant and hyperarousal symptoms correlated with DMN-CAN]

(continued)

Table 14.5 (continued)

6. Nicholson et al. (2020) [Alpha-rhythm EEG neurofeedback]
7. Shaw et al. (2023) [Alpha-rhythm EEG neurofeedback]
8. Schlumpf et al. (2021)
9. Lieberman et al. (2023)
10. Bell et al. (2019)
11. Gerin et al. (2016)
12. Kluetsch et al. (2014) [An early study which contains contradictory results for SN connectivity, and is now removed from the table]
13. Misaki et al. (2018)
14. Nicholson et al. (2018).
15. Lotfinia et al. (2023) [Review shows effectiveness of THC and Oxytocin
16. Norbury et al. (2021) [ketamine effect on RSN follows improvement]
17. Duek et al. (2023)
18. Philip et al. (2018) [PTSD & comorbid MDD patients]
19. Liston et al. (2014)
20. Zhao et al. (2023)
21 Oathes et al. (2024)
22. Zhang et al. (2024)
23 Sydnor et al. (2022)
24. Voigt et al. (2024) REVIEW (analysed on NF separately)

Table 14.6 Effects of different therapies on changes in functional activity and connectivity within and between different resting state networks and implications for treatment of PTSD

Therapeutic measure	Targeted networks	Effects	Implications for PTSD
Trauma-focused cognitive behaviour therapy[a]	DMN CEN	Improves DMN connectivity by reducing trauma-related rumination; enhances CEN activity for better cognitive control	Proven gold-standard therapy for PTSD with strong evidence of symptom reduction
Neurofeedback[b]	DMN SN	Restores DMN connectivity and reduces SN hyperactivity by training self-regulation of brain activity	Effective in normalizing network connectivity and reducing PTSD symptoms; remission rates ~61%
Mindfulness-based therapies[c]	DMN SN	Enhances DMN activity for self-referential processing; increases external connectivity of SN with DMN and DMN with CEN to lower emotional reactivity	Shown to improve emotional regulation and reduce PTSD symptoms through network balance restoration
Transcranial magnetic stimulation[c]	CEN SN	Decreases amygdala connectivity to CEN and sgACC to SN, CEN and DMN	Promising for targeting specific regions like the dorsolateral prefrontal cortex

CEN central executive network, *DMN* default mode network (DMN), *PTSD* posttraumatic stress disorder, *sgACC* subgenual anterior cingulate cortex, *SN* salience network

Sources:

[a] Table 14.3; Knaevelsrud and Maercker (2007); Bisson et al. (2013)

[b] Table 14.3; Nicholson et al. (2018)

[c] Table 14.3; Lanius et al. (2015)

References

Ahmadizadeh MJ, Rezaei M (2018) Unilateral right and bilateral dorsolateral prefrontal cortex transcranial magnetic stimulation in treatment post-traumatic stress disorder: a randomized controlled study. Brain Res Bull 140:334–340. https://doi.org/10.1016/j.brainresbull.2018.06.001

Akiki TJ, Averill CL, Abdallah CG (2017) A network-based neurobiological model of PTSD: evidence from structural and functional neuroimaging studies. Curr Psychiatry Rep 19(11):81. https://doi.org/10.1007/s11920-017-0840-4

Askovic M, Soh N, Elhindi J, Harris AWF (2023) Neurofeedback for post-traumatic stress disorder: systematic review and meta-analysis of clinical and neurophysiological outcomes. Eur J Psychotraumatol 14(2):2257435. https://doi.org/10.1080/20008066.2023.2257435

Bell AN, Moss D, Kallmeyer RJ (2019) Healing the neurophysiological roots of trauma: a controlled study examining LORETA z-score neurofeedback and HRV biofeedback for chronic PTSD. NeuroRegulation 6(2):54–70. https://doi.org/10.15540/nr.6.2.54

Berlim MT, Van Den Eynde F (2014) Repetitive transcranial magnetic stimulation over the dorsolateral prefrontal cortex for treating posttraumatic stress disorder: an exploratory meta-analysis of randomized, double-blind and sham-controlled trials. Can J Psychiatr 59(9):487–496. https://doi.org/10.1177/070674371405900905

Bisson JI, Roberts NP, Andrew M, Cooper R, Lewis C (2013) Psychological therapies for chronic post-traumatic stress disorder (PTSD) in adults. Cochrane Database Syst Rev 12:CD003388. https://doi.org/10.1002/14651858.CD003388.pub4

Bluhm RL, Williamson PC, Osuch EA, Frewen PA, Stevens TK, Boksman K, Neufeld RW, Théberge J, Lanius RA (2009) Alterations in default network connectivity in posttraumatic stress disorder related to early-life trauma. J Psychiatry Neurosci 34(3):187–194

Boggio PS et al (2010) Noninvasive brain stimulation with high-frequency and low-intensity repetitive transcranial magnetic stimulation treatment for posttraumatic stress disorder. J Clin Psychiatry 71(8):992–999. https://doi.org/10.4088/JCP.08m04638blu

Boyd JE, Lanius RA, McKinnon MC (2018) Mindfulness-based treatments for posttraumatic stress disorder: a review of the treatment literature and neurobiological evidence. J Psychiatry Neurosci 43(1):7–25. https://doi.org/10.1503/jpn.170021

Brown VM et al (2014) Altered resting-state functional connectivity of basolateral and centromedial amygdala complexes in posttraumatic stress disorder. Neuropsychopharmacology 39(2):351–359. https://doi.org/10.1038/npp.2013.197

Cash RFH et al (2019) Subgenual functional connectivity predicts antidepressant treatment response to transcranial magnetic stimulation: independent validation and evaluation of personalization. Biol Psychiatry 85(2):e5–e7. https://doi.org/10.1016/j.biopsych.2018.12.002

Chan A (2016) The fragmentation of self and others: the role of the default mode network in post-traumatic stress disorder. Pepperdine University Theses and Dissertations 660. https://digitalcommons.pepperdine.edu/etd/660

Chang CH, Liu CY, Chen SJ (2023) Accelerated theta-burst transcranial magnetic stimulation of the bilateral dorsolateral prefrontal cortex in a patient with post-traumatic stress disorder and major depressive disorder: case report. Front Psych 14:1061088. https://doi.org/10.3389/fpsyt.2023.1061088

Charquero-Ballester M et al (2022) Effective psychological therapy for PTSD changes the dynamics of specific large-scale brain networks. Hum Brain Mapp 43(10):3207–3220. https://doi.org/10.1002/hbm.25846

Chen AC, Etkin A (2013) Hippocampal network connectivity and activation differentiates post-traumatic stress disorder from generalized anxiety disorder. Neuropsychopharmacology 38(10):1889–1898. https://doi.org/10.1038/npp.2013.122

Cisler JM et al (2013) Neural processing correlates of assaultive violence exposure and PTSD symptoms during implicit threat processing: a network-level analysis among adolescent girls. Psychiatry Res Neuroimaging 214(3):238–246. https://doi.org/10.1016/j.pscychresns.2013.06.003

Cole J et al (2022) Efficacy of adjunctive D-cycloserine to intermittent theta-burst stimulation for major depressive disorder: a randomized clinical trial. JAMA Psychiatry 79(12):1153–1161. https://doi.org/10.1001/jamapsychiatry.2022.3255
Duek O et al (2023) Long term structural and functional neural changes following a single infusion of ketamine in PTSD. Neuropsychopharmacology 48(11):1648–1658. https://doi.org/10.1038/s41386-023-01606-3
Evans JW et al. (2018). Default mode connectivity in major depressive disorder measured up to 10 days after ketamine administration. Biol Psychiatry, 85(8), 582–590. https://doi.org/10.1016/j.biopsych.2018.01.027
Gerin MI et al (2016) Real-time fMRI neurofeedback with war veterans with chronic PTSD: a feasibility study. Front Psych 7:111. https://doi.org/10.3389/fpsyt.2016.00111
Graham BM, Milad MR (2011) The study of fear extinction: implications for anxiety disorders. Am J Psychiatry 168(12):1255–1265. https://doi.org/10.1176/appi.ajp.2011.11040557
Hajong D (2021) The importance of the default mode network (DMN) in post-traumatic stress disorder (PTSD). Int J Clin Health Sci. https://icjs.us/the-importance-of-the-default-mode-network-dmn-in-post-traumatic-stress-disorder-ptsd. Accessed: 28 Oct 2025
Hayes JP, Vanelzakker MB, Shin LM (2012) Emotion and cognition interactions in PTSD: a review of neurocognitive and neuroimaging studies. Front Integr Neurosci 6:89. https://doi.org/10.3389/fnint.2012.00089
Joyce MKP, García-Cabezas MÁ, John YJ, Barbas H (2020) Serial prefrontal pathways are positioned to balance cognition and emotion in primates. J Neurosci 40(43):8306–8328. https://doi.org/10.1523/JNEUROSCI.0860-20.2020
Kearney-Ramos TE et al (2018) Transdiagnostic effects of ventromedial prefrontal cortex transcranial magnetic stimulation on cue reactivity. Biol Psychiatry Cogn Neurosci Neuroimaging 3(7):599–609. https://doi.org/10.1016/j.bpsc.2018.03.016
King AP et al (2016) A pilot study of mindfulness-based exposure therapy in OEF/OIF combat veterans with PTSD: altered medial frontal cortex and amygdala responses in social-emotional processing. Front Psych 7:154. https://doi.org/10.3389/fpsyt.2016.00154
Kluetsch RC, Ros T, Théberge J, Frewen PA, Calhoun VD, Schmahl C, Jetly R, Lanius RA (2014) Plastic modulation of PTSD resting-state networks and subjective wellbeing by EEG neurofeedback. Acta Psychiatr Scand 130(2):123–136. https://doi.org/10.1111/acps.12229
Knaevelsrud C, Maercker A (2007) Internet-based treatment for PTSD reduces distress and facilitates the development of a strong therapeutic alliance: a randomized controlled clinical trial. BMC Psychiatry 7:13. https://doi.org/10.1186/1471-244X-7-13
Koch SB et al (2016) Aberrant resting-state brain activity in posttraumatic stress disorder: a meta-analysis and systematic review. Depress Anxiety 33(7):592–605. https://doi.org/10.1002/da.22478
Koek RJ et al (2019) Neuromodulatory treatments for post-traumatic stress disorder (PTSD). Prog Neuro-Psychopharmacol Biol Psychiatry 89:148–160. https://doi.org/10.1016/j.pnpbp.2019.01.004
Korgaonkar MS et al (2020) Intrinsic connectomes underlying response to trauma-focused psychotherapy in post-traumatic stress disorder. Transl Psychiatry 10:1270. https://doi.org/10.1038/s41398-020-00938-8
Lanius RA et al (2015) Restoring large-scale brain networks in PTSD and related disorders: a proposal for neuroscientifically informed treatment interventions. Eur J Psychotraumatol 6:27313. https://doi.org/10.3402/ejpt.v6.27313
Lieberman JM et al (2023) A tale of two targets: examining the differential effects of posterior cingulate cortex- and amygdala-targeted fMRI-neurofeedback in a PTSD pilot study. Front Neurosci 17:1229729. https://doi.org/10.3389/fnins.2023.1229729
Liston C et al (2014) Default mode network mechanisms of transcranial magnetic stimulation in depression. Biol Psychiatry 76(7):517–526. https://doi.org/10.1016/j.biopsych.2014.01.023
Lotfinia S, Afshar A, Yaseri A, Olff M, Quidé Y (2023) Functional brain changes after alternative pharmacological interventions in posttraumatic stress disorder: a systematic review of clinical trials. Brain Behav 13(12):e3292. https://doi.org/10.1002/brb3.3292

Makale MT, Abbasi S, Nybo C, Keifer J, Christman L, Fairchild JK, Yesavage J, Blum K, Gold MS, Baron D, Cadet JL, Elman I, Dennen CA, Murphy KT (2023) Personalized repetitive transcranial magnetic stimulation (prtms®) for post-traumatic stress disorder in military combat veterans. Heliyon 9(8):e18943. https://doi.org/10.1016/j.heliyon.2023.e18943

McDonald WM, van Rooij SJH (2019) Targeting PTSD. Am J Psychiatry 176(11):894–896. https://doi.org/10.1176/appi.ajp.2019.19080895

McGirr A et al (2021) Repetitive transcranial magnetic stimulation for the treatment of post-traumatic stress disorder: a systematic review and network meta-analysis. Can J Psychiatr 66(9):763–773. https://doi.org/10.1177/0706743720982432

Morriss R, Briley PM, Webster L, Abdelghani M, Barber S, Bates P, Brookes C, Hall B, Ingram L, Kurkar M, Lankappa S, Liddle PF, McAllister-Williams RH, O'Neil-Kerr A, Pszczolkowski S, Suazo Di Paola A, Walters Y, Auer DP (2024) Connectivity-guided intermittent theta burst versus repetitive transcranial magnetic stimulation for treatment-resistant depression: a randomized controlled trial. Nat Med 30(2):403–413. https://doi.org/10.1038/s41591-023-02764-z

Nicholson AA, Ros T, Frewen PA, Densmore M, Théberge J, Kluetsch RC, Jetly R, Lanius RA (2016) Alpha oscillation neurofeedback modulates amygdala complex connectivity and arousal in posttraumatic stress disorder. Neuroimage Clin 12:506–516. https://doi.org/10.1016/j.nicl.2016.07.006

Nicholson AA et al (2018) Intrinsic connectivity network dynamics in PTSD during amygdala downregulation using real-time fMRI neurofeedback: a preliminary analysis. Hum Brain Mapp 39(11):4258–4275. https://doi.org/10.1002/hbm.24244

Norbury A et al (2021) Neuroimaging correlates and predictors of response to repeated-dose intravenous ketamine in PTSD: preliminary evidence. Neuropsychopharmacology 46(13):2266–2277. https://doi.org/10.1038/s41386-021-01104-4

Oathes DJ et al (2024) Clinical response to fMRI-guided compared to non-image-guided rTMS in depression and PTSD: a randomized trial. medRxiv [preprint]. https://doi.org/10.1101/2024.07.29.24311191

Patel R, Spreng RN, Shin LM, Girard TA (2012) Neurocircuitry models of posttraumatic stress disorder and beyond: a meta-analysis of functional neuroimaging studies. Neurosci Biobehav Rev 36(9):2130–2142. https://doi.org/10.1016/j.neubiorev.2012.06.003

Philip NS et al (2018) Network mechanisms of clinical response to transcranial magnetic stimulation in posttraumatic stress disorder and major depressive disorder. Biol Psychiatry 83(4):263–272. https://doi.org/10.1016/j.biopsych.2017.07.021

Philip NS, Aiken EE, Kelley ME, Burch W, Waterman L, Holtzheimer PE (2019) Synchronized transcranial magnetic stimulation for posttraumatic stress disorder and comorbid major depression. Brain Stimul 12(5):1335–1337. https://doi.org/10.1016/j.brs.2019.06.010

Pitman RK, Rasmusson AM, Koenen KC, Shin LM, Orr SP, Gilbertson MW, Milad MR, Liberzon I (2012) Biological studies of post-traumatic stress disorder. Nat Rev Neurosci 13(11):769–787. https://doi.org/10.1038/nrn3339

Rabellino D, Tursich M, Frewen PA, Daniels JK, Densmore M, Théberge J, Lanius RA (2015) Intrinsic connectivity networks in post-traumatic stress disorder during sub- and supraliminal processing of threat-related stimuli. Acta Psychiatr Scand 132(5):365–378. https://doi.org/10.1111/acps.12418

Rabinak CA, Angstadt M, Welsh RC, Kennedy AE, Lyubkin M, Martis B, Phan KL (2011) Altered amygdala resting-state functional connectivity in post-traumatic stress disorder. Front Psychiatry 2:62. https://doi.org/10.3389/fpsyt.2011.00062

Shaw SB, Nicholson AA, Ros T, Harricharan S, Terpou B, Densmore M, Théberge J, Frewen P, Lanius RA (2023) Increased top-down control of emotions during symptom provocation working memory tasks following a randomized controlled trial of alpha-down neurofeedback in PTSD. Neuroimage Clin 37:103313. https://doi.org/10.1016/j.nicl.2023.103313

Shin LM, Liberzon I (2010) The neurocircuitry of fear, stress, and anxiety disorders. Neuropsychopharmacology 35(1):169–191. https://doi.org/10.1038/npp.2009.83

Siddiqi SH et al (2024) A potential target for noninvasive neuromodulation of PTSD symptoms derived from focal brain lesions in veterans. Nat Neurosci 27(11):2231–2239. https://doi.org/10.1038/s41593-024-01772-7

Sripada RK, King AP, Welsh RC, Garfinkel SN, Wang X, Sripada CS, Liberzon I (2012) Neural dysregulation in posttraumatic stress disorder: evidence for disrupted equilibrium between salience and default mode brain networks. Psychosom Med 74(9):904–911. https://doi.org/10.1097/PSY.0b013e318273bf33

St Jacques PL, Kragel PA, Rubin DC (2013) Neural networks supporting autobiographical memory retrieval in posttraumatic stress disorder. Cogn Affect Behav Neurosci 13(3):554–566. https://doi.org/10.3758/s13415-013-0157-7

Taghva A, Silvetz R, Ring A, Kim KY, Murphy KT, Liu CY, Jin Y (2015) Magnetic resonance therapy improves clinical phenotype and EEG alpha power in posttraumatic stress disorder. Trauma Mon 20(4):e27360. https://doi.org/10.5812/traumamon.27360

Vasavada MM, Loureiro J, Kubicki A, Sahib A, Wade B, Hellemann G, Espinoza RT, Congdon E, Narr KL, Leaver AM (2021) Effects of serial ketamine infusions on corticolimbic functional connectivity in major depression. Biol Psychiatry Cogn Neurosci Neuroimaging 6(7):735–744. https://doi.org/10.1016/j.bpsc.2020.06.015

Viard A, Mutlu J, Chanraud S, Guenolé F, Egler PJ, Gérardin P, Baleyte JM, Dayan J, Eustache F, Guillery-Girard B (2019) Altered default mode network connectivity in adolescents with post-traumatic stress disorder. Neuroimage Clin 22:101731. https://doi.org/10.1016/j.nicl.2019.101731

Voigt JD, Mosier M, Tendler A (2024) Systematic review and meta-analysis of neurofeedback and its effect on posttraumatic stress disorder. Front Psychiatry 15:1323485. https://doi.org/10.3389/fpsyt.2024.1323485

Vuper TC, Philippi CL, Bruce SE (2021) Altered resting-state functional connectivity of the default mode and central executive networks following cognitive processing therapy for PTSD. Behav Brain Res 409:113312. https://doi.org/10.1016/j.bbr.2021.113312

Wang T et al (2016) Altered resting-state functional activity in posttraumatic stress disorder: a quantitative meta-analysis. Sci Rep 6:27131. https://doi.org/10.1038/srep27131

Woollams AM, Lindley LJ, Pobric G, Hoffman P (2017) Laterality of anterior temporal lobe repetitive transcranial magnetic stimulation determines the degree of disruption in picture naming. Brain Struct Funct 222(8):3749–3759. https://doi.org/10.1007/s00429-017-1430-2

Zandvakili A, Barredo J, Swearingen HR, Aiken EM, Berlow YA, Greenberg BD, Carpenter LL, Philip NS (2020) Mapping PTSD symptoms to brain networks: a machine learning study. Transl Psychiatry 10(1):195. https://doi.org/10.1038/s41398-020-00879-2

Zhang Y et al (2015) Intranetwork and internetwork functional connectivity alterations in post-traumatic stress disorder. J Affect Disord 187:114–121. https://doi.org/10.1016/j.jad.2015.08.043

Zhang Y et al (2024) Efficacy of MRI-guided rTMS for post-traumatic stress disorder by modulating amygdala activity: study protocol for a randomised controlled trial. BMJ Open 14(7):e081751. https://doi.org/10.1136/bmjopen-2023-081751

Zhao Z et al (2023) Amygdala downregulation training using fMRI neurofeedback in post-traumatic stress disorder: a randomized, double-blind trial. Transl Psychiatry 13:1177. https://doi.org/10.1038/s41398-023-02467-6

Chapter 15
Abnormalities in Resting State Networks in PTSD and Treatment with Electroencephalography (EEG) and Neurofeedback Interventions

15.1 Electroencephalography (EEG) Methodology

Electroencephalography (EEG) signals are generated by the integration of neural activity at multiple spatial and temporal scales (Fig. 15.1). The spatial scales cover neurons, local networks of thousands of neurons and the entire brain regions of millions of neurons. Recorded at these different spatial scales, neuronal activities run at specific time scales in the form of spikes, local field potential (largely derived from the summation of dendritic activities over small groups of neurons) or EEG recordings.[1]

Each electrode placement site (Fig. 15.2) has a letter to identify the lobe, or area, of the brain it is reading from: prefrontal (Fp), frontal (F), temporal (T), parietal (P), occipital (O) and central (C). Note that there is no 'central lobe'; due to their placement, and depending on the individual, the 'C' electrodes can exhibit EEG activity more typical of frontal, temporal and some parietal–occipital activity.

EEG signals are categorized into distinct frequency bands, including delta, theta, alpha and beta waves (Fig. 15.3). As alpha-wave power is thought to reflect the inverse of cortical activity, right minus left hemisphere alpha power produces a metric suggestive of relatively greater left than right hemisphere cortical activity.[2]

[1] Le Van Quyen (2011).

[2] Butt et al. (2019).

M. Bennett, *Veteran's Stress, Trauma and Synapse Rehabilitation*, https://doi.org/10.1007/978-981-95-9594-5_15

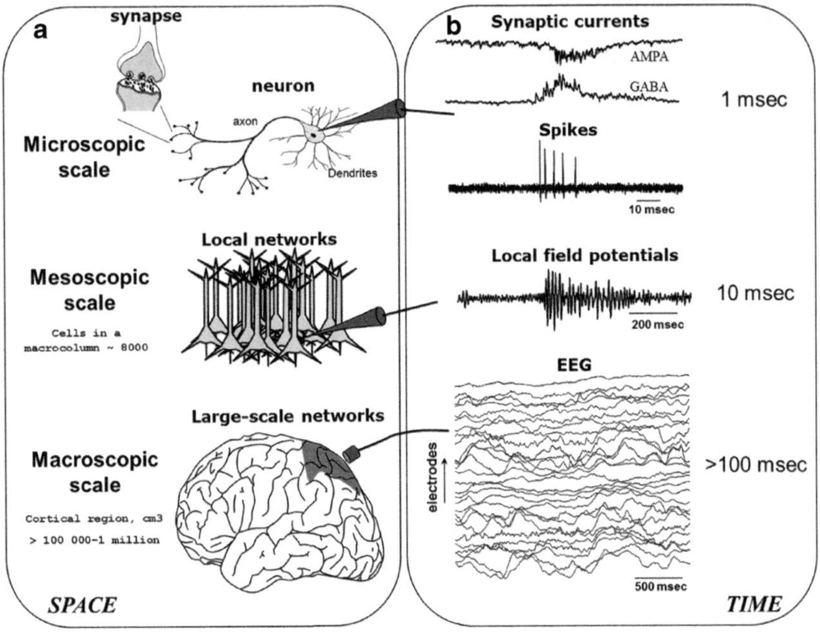

Fig. 15.1 The generation of electroencephalogram (EEG) network oscillations in (**a**) space and (**b**) time scales. (Reproduced from Le Van Quyen (2011) with permission from Elsevier)

15.2 EEG Alpha-Wave Power, PTSD and Neurofeedback

Posttraumatic stress disorder (PTSD) symptoms are correlated with peak alpha-wave power.[3] There is a highly significant negative correlation between frontotemporal alpha-wave power and PTSD.[4] EEG-based neurofeedback (EEG-NF) can be used to modulate the alpha-wave power of a PTSD patient and thus ameliorate their symptoms. An excellent example of this is provided by a video-game training technique. The patients were asked to maintain an astronaut on a track while avoiding dangerous hazards, with the position of the astronaut linked to each patient's alpha-wave power to indicate whether it was above or below a threshold related to eventual suppression of PTSD, with patients receiving a reward for staying on track.[5] At the end of the training period, the neurofeedback led to a decrease in alpha-wave power, despite an increase in this power being correlated with a decrease of PTSD symptoms. However, following the training period, there was a persistent 'rebound' in alpha-wave power and amelioration of PTSD symptoms (Fig. 15.4).[6]

[3] Lobo et al. (2015).

[4] Peddi et al. (2024).

[5] See du Bois et al. (2021).

[6] Kluetsch et al. (2014); see also Reiter et al. (2016).

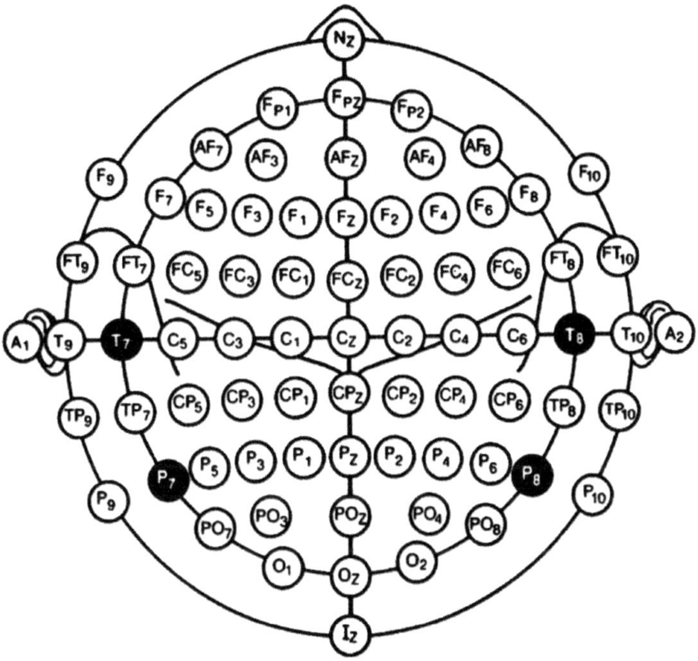

Fig. 15.2 Electroencephalography electrode placements

The earliest use of EEG alpha-wave-centred neurofeedback, in 2016, showed that PTSD patients provided with this treatment showed a highly significant drop in their clinician-administered PTSD scale (CAPS) score (from an average of 66 to 42, i.e. 36%; $d = -2.23$) for the amelioration of PTSD (Fig. 15.5).[7] This is not much different to the most recent meta-analysis, which shows an average reduction in CAPS score of 31%.[8] Theta-wave power can also be used in EEG-NF to ameliorate PTSD.[9]

[7] Van der Kolk et al. (2016); see also Ros et al. (2017).

[8] Voigt et al. (2024).

[9] Noohi et al. (2017).

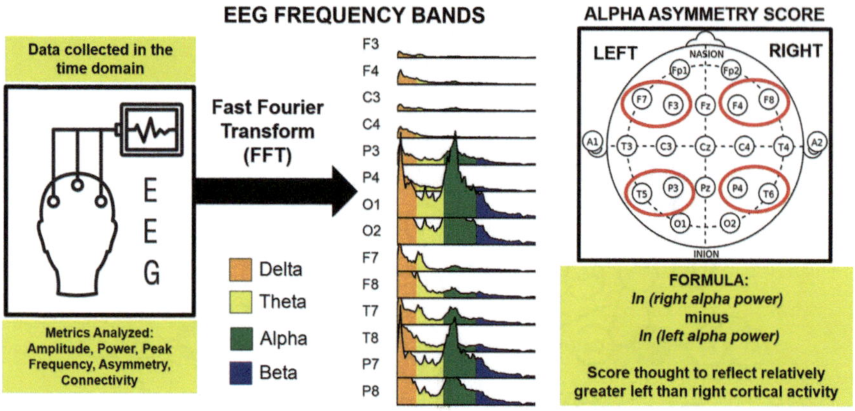

Fig. 15.3 Electroencephalography (EEG) data collected in the time domain and run through a fast Fourier transform (FFT) to evaluate ranges of frequencies, or bands, including delta, theta, alpha and beta bands. Alpha asymmetry scores are typically computed for frontal (F7, F3, F4 and F8) and temporal/parietal (T5, P3, P4, T6) EEG electrodes measured at the scalp. (Reproduced from Butt et al. (2019). Creative Commons CC BY 4.0 (https://creativecommons.org/licenses/by/4.0/))

15.3 Increasing EEG Alpha Power in PTSD Patients Is Accompanied by Normalization of Functional Activity and Connectivity in Resting State Networks

15.3.1 Default Mode Network (DMN) Functional Activity

In a recent major review of the effects of neurofeedback on patients with PTSD, it was shown that a significant reduced alpha-wave (8–12 Hz) power occurs in their EEG.[10] This is especially so in the anterior components of the default mode network (DMN) with this rectified using neurofeedback rebound of the alpha-wave power, leading to remission of their PTSD with a rate of 60% at a 3-month follow-up (Fig. 15.5).

The alpha-wave power of the EEG is correlated with the DMN activity as determined by functional magnetic resonance imaging (fMRI) and EEG research. Thus, the low functional activity of the DMN in PTSD, especially as observed in the medial prefrontal cortex (mPFC) and the posterior cingulate cortex (PCC) hubs of the DMN (see Chap. 14), is offset by neurofeedback enhancement of alpha-wave power.

[10] Nicholson et al. (2023).

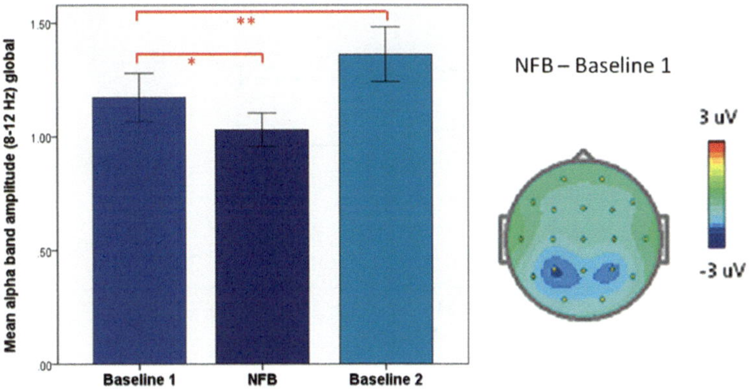

Fig. 15.4 Homeostatic "rebound" following desynchronizing neurofeedback (NFB). Left: mean (SEM) global alpha amplitude in PTSD patients: before (baseline1), during (neurofeedback) and right after neurofeedback (baseline2). Right: topographic plot of mean alpha amplitude change during neurofeedback (NFB), relative to resting state (baseline1).*p < 0.05, ** P < 0.005. (Reproduced from Kluetsch et al. (2014) with permission from Wiley)

15.3.2 Default Mode Network (DMN) Functional Connectivity

PTSD symptoms are highly dependent on activity in the amygdala, many nuclei of which are functionally connected within the DMN and the salience network (SN). After EEG-NF targeting alpha-wave power centred on the PCC/precuneus component of the DMN for 30 min, there was most significantly an increase in functional activity between the amygdala and mPFC associated with decreased PTSD symptoms (Fig. 15.6).[11]

In patients with PTSD receiving alpha-wave power EEG-NF, 3 months after the treatment, there was a change in the functional connectivity of the DMN and the SN towards the normal, accompanied by a significant decrease in PTSD symptoms of 50% compared with the control group.[12]

15.3.3 DMN Intrinsic Functional Connectivity

Even a single session of alpha-wave-based EEG-NF presented in such a way as to provide a rebound increase in alpha-wave power restores functional connectivity within the DMN (mPFC, PCC) of PTSD patients.[13]

[11] Nicholson et al. (2016).

[12] Nicholson et al. (2020).

[13] Kluetsch et al. (2014).

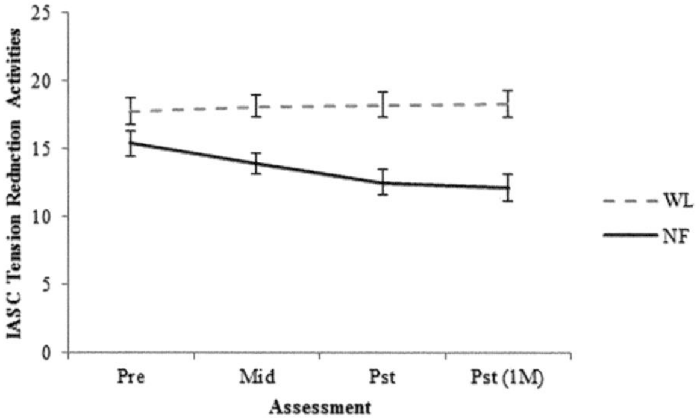

Fig. 15.5 Change in posttraumatic stress disorder (PTSD) symptom severity (total CAPS score) as a function of treatment condition, showing standard error bars included at each assessment. CAPS, clinician-administered PTSD scale; IASC, Inventory of Altered Self-Capacities; M, month; Mid, mid-treatment assessment; NF, patients receiving neurofeedback treatment; Pre, pre-treatment assessment; Pst, post-treatment assessment; WL, waitlist patients (no treatment). (Reproduced from Van der Kolk et al. (2016), Creative Commons CC BY 4.0 (https://creativecommons.org/licenses/by/4.0/))

15.3.4 DMN Extrinsic Functional Connectivity

Alpha-wave power enhancement using EEG-NF to normalize the functional activity and connectivity between the central executive network (CEN) (dorsolateral prefrontal cortex) and the DMN (anterior and posterior parts) reduces PTSD symptoms as indicated by the CAPS score by over 40% at 3-month follow-up.[14]

Abnormalities in DMN and SN functional connectivity in PTSD are corrected with alpha-wave-based EEG-NF, leading to a 50% remission rate in PTSD.[15]

15.4 fMRI Measures of Functional Activity and Connectivity Modulated by Neurofeedback

15.4.1 Amygdala-Centred fMRI Activity

Most studies using fMRI measured blood-oxygenation-level–dependent (BOLD) signals as an indicator of functional activity to be regulated in a neurofeedback paradigm have naturally centred on amygdala activity, which is closely aligned with PTSD. PTSD patients have been taught to control their heightened amygdala activity when confronted with personalized trauma scripts, leading to an average drop in

[14] Shaw et al. (2023).

[15] Nicholson et al. (2020).

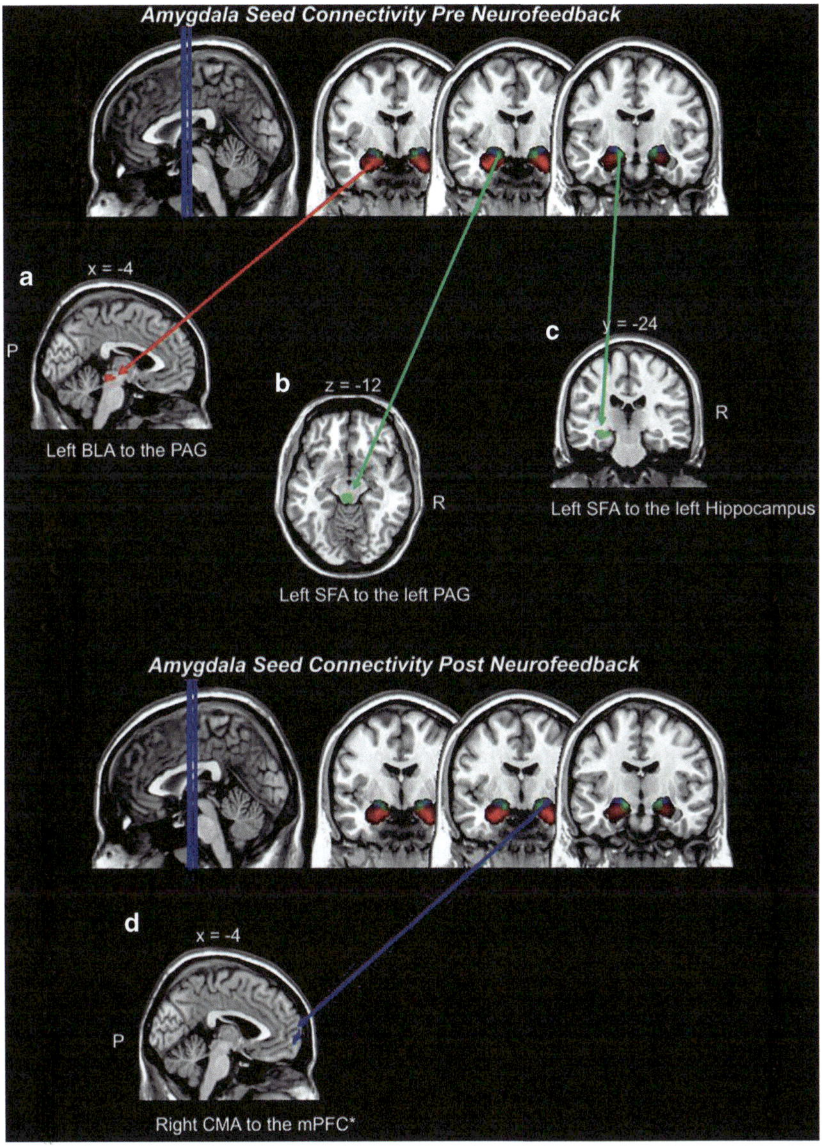

Fig. 15.6 Clusters representing greater connectivity from amygdala complex seeds before neurofeedback intervention (abbreviations defined below): (**a**) left BLA to the PAG; (**b**) left SFA to the left PAG; (**c**) left SFA to the left hippocampus, as compared to after neurofeedback within PTSD patients. Non-significant differences were found with respect to the right BLA, right SFA and bilateral CMA, when examining increased amygdala functional connectivity pre-neurofeedback as compared to post-neurofeedback. Inversely, increased amygdala complex connectivity occurred after neurofeedback, as compared to before neurofeedback. (**d**) Right CMA to the mPFC. Here, non-significant differences were found when examining increased amygdala complex connectivity post-neurofeedback, as compared to pre-neurofeedback, for the left CMA, bilateral BLA and bilateral SFA. X, Y and Z indicate the position of brain slices. R, right hemisphere; P, posterior; BLA, basolateral amygdala complex; CMA, centro-medial amygdala complex; SFA, superficial amygdala complex; mPFC, medial prefrontal cortex; PAG, periaqueductal grey. (Reproduced from Nicholson et al. (2016) with permission from Elsevier. Creative Commons CC BY 4.0 (https://creativecommons.org/licenses/by/4.0/))

their CAPS score of about 30, and one in three achieving remission.[16] Amelioration of PTSD symptoms in combat veterans was also achieved, as was functional connectivity, using the activity of the amygdala in the feedback signal.[17] There is good evidence then for lessening PTSD symptoms using fMRI-NF on the amygdala.[18]

15.4.2 Posterior Cingulate Cortex–Centred fMRI Activity

It has recently been proposed that PTSD is best ameliorated using fMRI measures of functional activity in the PCC, which is raised in PTSD, and can be lowered using neurofeedback based on this activity, with PTSD symptoms also declining.[19]

15.4.3 Comparison Between EEG-NF and fMRI-NF

Comparisons have been made between using functional activity determined for the amygdala with fMRI as the neurofeedback signal compared with those using alpha-wave power in the EEG as the signal in attempts to ameliorate PTSD. A meta-analysis found that the fMRI approach has been shown to be considerably less effective than the EEG approach, with EEG-NF giving an average Hedge's g of −1.13 compared with that for fMRI-NF of −0.37 (based on amygdala activity).[20] The authors of the meta-analysis further found that:

> EEG was more effective than fMRI for PTSD symptoms, and the effect on PTSD symptoms was higher than on anxiety and depression. There was no difference in the effectiveness of the training sessions. Our findings showed that EEG-based neuro-feedback training was more helpful for training PTSD symptoms. Additionally, the methods were also shown to be valid for evaluating clinical PTSD diagnoses.[21]

---

[16] Gerin et al. (2016). See also Nicholson et al. (2018) and Zotev et al. (2018).

[17] Misaki et al. (2018).

[18] Reviewed in Tursic et al. (2020).

[19] Lieberman et al. (2023).

[20] Hong and Park (2022).

[21] Hong and Park (2022) abstract.

References

Butt M, Espinal E, Aupperle RL, Nikulina V, Stewart JL (2019) The electrical aftermath: brain signals of posttraumatic stress disorder filtered through a clinical lens. Front Psych 10:368. https://doi.org/10.3389/fpsyt.2019.00368

du Bois N, Bigirimana AD, Korik A, Kéthina LG, Rutembesa E, Mutabaruka J, Mutesa L, Prasad G, Jansen S, Coyle DH (2021) Neurofeedback with low-cost, wearable electroencephalography (EEG) reduces symptoms in chronic post-traumatic stress disorder. J Affect Disord 295:1319–1334. https://doi.org/10.1016/j.jad.2021.08.071

Gerin MI, Fichtenholtz H, Roy A, Walsh CJ, Krystal JH, Southwick S, Hampson M (2016) Real-time fMRI neurofeedback with war veterans with chronic PTSD: a feasibility study. Front Psych 7:111. https://doi.org/10.3389/fpsyt.2016.00111

Hong J, Park JH (2022) Efficacy of neurofeedback training for PTSD symptoms: a systematic review and meta-analysis. Int J Environ Res Public Health 19(20):13096. https://doi.org/10.3390/ijerph192013096

Kluetsch RC, Ros T, Théberge J, Frewen PA, Calhoun VD, Schmahl C, Jetly R, Lanius RA (2014) Plastic modulation of PTSD resting-state networks and subjective wellbeing by EEG neurofeedback. Acta Psychiatr Scand 130(2):123–136. https://doi.org/10.1111/acps.12229

Le Van Quyen M (2011) The brainweb of cross-scale interactions. New Ideas Psychol 29:57–63. https://doi.org/10.1016/j.newideapsych.2010.10.002

Lieberman JM, Rabellino D, Densmore M, Frewen PA, Steyrl D, Scharnowski F, Théberge J, Hosseini-Kamkar N, Neufeld RWJ, Jetly R, Frey BN, Ros T, Lanius RA, Nicholson AA (2023) A tale of two targets: examining the differential effects of posterior cingulate cortex- and amygdala-targeted fMRI neurofeedback in a PTSD pilot study. Front Neurosci 17:1229729. https://doi.org/10.3389/fnins.2023.1229729

Lobo I, Portugal LC, Figueira I, Volchan E, David I, Garcia Pereira M, de Oliveira L (2015) EEG correlates of the severity of posttraumatic stress symptoms: a systematic review of the dimensional PTSD literature. J Affect Disord 183:210–220. https://doi.org/10.1016/j.jad.2015.05.015

Misaki M, Phillips R, Zotev V, Wong CK, Wurfel BE, Krueger F, Feldner M, Bodurka J (2018) Real-time fMRI amygdala neurofeedback positive emotional training normalized resting-state functional connectivity in combat veterans with and without PTSD: a connectome-wide investigation. Neuroimage Clin 20:543–555. https://doi.org/10.1016/j.nicl.2018.08.025

Nicholson AA, Ros T, Frewen PA, Densmore M, Théberge J, Kluetsch RC, Jetly R, Lanius RA (2016) Alpha oscillation neurofeedback modulates amygdala complex connectivity and arousal in posttraumatic stress disorder. Neuroimage Clin 12:506–516. https://doi.org/10.1016/j.nicl.2016.07.006

Nicholson AA, Rabellino D, Densmore M, Frewen PA, Paret C, Kluetsch R, Schmahl C, Théberge J, Ros T, Neufeld RWJ, McKinnon MC, Reiss JP, Jetly R, Lanius RA (2018) Intrinsic connectivity network dynamics in PTSD during amygdala downregulation using real-time fMRI neurofeedback: a preliminary analysis. Hum Brain Mapp 39(11):4258–4275. https://doi.org/10.1002/hbm.24244

Nicholson AA, Harricharan S, Densmore M, Neufeld RWJ, Ros T, McKinnon MC, Frewen PA, Théberge J, Jetly R, Pedlar D, Lanius RA (2020) Classifying heterogeneous presentations of PTSD via the default mode, central executive, and salience networks with machine learning. Neuroimage Clin 27:102262. https://doi.org/10.1016/j.nicl.2020.102262

Nicholson AA, Densmore M, Frewen PA, Neufeld RWJ, Théberge J, Jetly R, Lanius RA, Ros T (2023) Homeostatic normalization of alpha brain rhythms within the default-mode network and reduced symptoms in post-traumatic stress disorder following a randomized controlled trial of electroencephalogram neurofeedback. Brain Commun 5(2):fcad068. https://doi.org/10.1093/braincomms/fcad068

Noohi S, Miraghaie AM, Arabi A, Nooripour R (2017) Effectiveness of neurofeedback treatment with alpha/theta method on PTSD symptoms and their executive function. Biomed Res 28(5):2019–2027

Peddi A, Sendi MSE, Minton ST, Hinojosa CA, West E, Langhinrichsen-Rohling R, Ressler KJ, Calhoun VD, van Rooij SJH (2024) Towards predicting PTSD symptom severity using portable EEG-derived biomarkers. medRxiv [Preprint]:2024.07.17.24310570. https://doi.org/10.1101/2024.07.17.24310570

Reiter K, Andersen SB, Carlsson J (2016) Neurofeedback treatment and posttraumatic stress disorder: effectiveness of neurofeedback on posttraumatic stress disorder and the optimal choice of protocol. J Nerv Ment Dis 204(2):69–77. https://doi.org/10.1097/NMD.0000000000000418

Ros T, Frewen P, Théberge J, Michela A, Kluetsch R, Mueller A, Candrian G, Jetly R, Vuilleumier P, Lanius RA (2017) Neurofeedback tunes scale-free dynamics in spontaneous brain activity. Cereb Cortex 27(10):4911–4922. https://doi.org/10.1093/cercor/bhw285

Shaw SB, Nicholson AA, Ros T, Harricharan S, Terpou B, Densmore M, Théberge J, Frewen P, Lanius RA (2023) Increased top-down control of emotions during symptom provocation working memory tasks following a randomized controlled trial of alpha-down neurofeedback in PTSD. Neuroimage Clin 37:103313. https://doi.org/10.1016/j.nicl.2023.103313

Tursic A, Eck J, Lührs M, Linden DEJ, Goebel R (2020) A systematic review of fMRI neurofeedback reporting and effects in clinical populations. Neuroimage Clin 28:102496. https://doi.org/10.1016/j.nicl.2020.102496

van der Kolk BA, Hodgdon H, Gapen M, Musicaro R, Suvak MK, Hamlin E, Spinazzola J (2016) A randomized controlled study of neurofeedback for chronic PTSD. PLoS One 11(12):e0166752. https://doi.org/10.1371/journal.pone.0166752

Voigt JD, Mosier M, Tendler A (2024) Systematic review and meta-analysis of neurofeedback and its effect on posttraumatic stress disorder. Front Psych 15:1323485. https://doi.org/10.3389/fpsyt.2024.1323485

Zotev V, Phillips R, Misaki M, Wong CK, Wurfel BE, Krueger F, Feldner M, Bodurka J (2018) Real-time fMRI neurofeedback training of the amygdala activity with simultaneous EEG in veterans with combat-related PTSD. Neuroimage Clin 19:106–121. https://doi.org/10.1016/j.nicl.2018.04.010

Part VI
Comparison of Interventions for Treatment of PTSD

Chapter 16
Comparison of Psychotherapies, Pharmacotherapies and Neurotherapies for PTSD

16.1 Most Effective Psychotherapies and Pharmacotherapies

Table 16.1 summarizes effect sizes from three meta-analyses of psychotherapies and pharmacotherapies on alleviating posttraumatic stress disorder (PTSD) symptoms overall, including one meta-analysis that also looked at effects on intrusion (B) and avoidance (C) symptom clusters (see Sect. 2.1). A 1998 meta-analysis concluded that eye movement desensitization and reprocessing (EMDR) and cognitive behaviour therapy (CBT) were most effective among psychotherapies, while selective serotonin reuptake inhibitors (SSRIs), such as paroxetine, had the greatest effect size among pharmacotherapies.[1] The study showed that EMDR was more effective than either CBT or SSRIs in treating intrusions and avoidance symptom clusters. A 2013 meta-analysis concluded that cognitive therapy (CT) has a high effect size among psychotherapies, followed by EMDR and prolonged exposure (PE) therapy, with pharmacotherapies being less effective.[2] More recently, a 2023 meta-analysis, which only looked at psychotherapies, concluded that EMDR is the best therapy overall.[3]

[1] Van Etten and Taylor (1998).

[2] Watts et al. (2013).

[3] Yunitri et al. (2023).

M. Bennett, *Veteran's Stress, Trauma and Synapse Rehabilitation*,
https://doi.org/10.1007/978-981-95-9594-5_16

Table 16.1 Results of meta-analyses comparing best effect sizes of treatments for posttraumatic stress disorder (PTSD) symptoms

Therapy	PTSD overall	Intrusions (B)	Avoidance (C)
Eye movement desensitization and reprocessing therapy	2.27[a], 1.39[b], 1.01[c]	2.07[a]	2.34[a]
Cognitive behaviour therapy	1.93[a]	1.47[a]	1.45[a]
Cognitive therapy	1.63[c], 1.33[b]	–	–
Prolonged exposure therapy	1.10[b], 1.08[c]	–	–
Selective serotonin reuptake inhibitors	1.43[a], 0.74[c]	1.28[a]	1.32[a]

Notes:
[a]Effect size is Cohen's d: Van Etten and Taylor (1998), Tables 2 and 4
[b]Yunitri et al. (2023), Table 4
[c]Effect size is Hedge's g: Watts et al. (2013)

16.2 Comparison of Treatments for PTSD According to National and International Guidelines

A comparison has been made between 14 national and international guidelines for the treatment of PTSDs, including from Australia, Canada, South Africa, the United Kingdom, and the United States.[4] (When comparing treatment guidelines from different countries, it must be remembered that trauma-focused CBT (TF-CBT) for adults in the United Kingdom and Europe includes psychotherapies, such as CT, PE and cognitive processing therapy (CPT), whereas in the USA TF-CBT (for adolescents and children), PE, CPT and CT are regarded as individual therapies.) The results of this comparison, and the recommendations in the latest Australian national guidelines and the American Psychological Association (APA) guidelines, are shown in Table 16.2.

Table 16.2 shows that psychotherapies, such as CPT, CBT and TF-CBT, and PE are first-line therapies in all three sets of guidelines. EMDR is a first-line therapy in Australia and 6 of the 14 compared guidelines, but has only conditional recommendation as a second-line therapy in the APA guidelines. Narrative exposure therapy is a conditional, second-line therapy in the Australian and APA guidelines. For pharmacotherapies, SSRIs and serotonin–norepinephrine reuptake inhibitors (SNRIs) are conditional, second-line interventions in the Australian and APA guidelines; however, among the 14 compared guidelines, SSRIs are first-line interventions in 13 of them, while the SNRI venlafaxine is a first-line intervention in ten of them. All guidelines exclude ketamine and 3,4-methylenedioxymethamphetamine (MDMA) from recommended treatments. (Similarly, transcranial magnetic stimulation (TMS) and electroencephalography (EEG)-guided neurofeedback do not appear in any PTSD guidelines as they are regarded as 'experimental'.)

Table 16.3 sets out US Food and Drug Administration (FDA) approval status of neurotherapies and pharmacotherapies (but not psychotherapies as they are not subject to the FDA).

[4]Martin et al. (2021).

Table 16.2 Recommended interventions according to guidelines for treatment of PTSD

Intervention	Context	Australian guidelines[a]	Compared guidelines[b]	APA guidelines[c]
CPT	12-session manualized treatment; addresses trauma themes (safety, trust, etc.)	Strong (first-line)	Strong (first-line)	Strong (first-line)
TF-CBT	Integrates CBT with trauma processing; psychoeducation, exposure, cognitive restructuring for adults	Strong (first-line)	Strong (first-line)	Strong (first-line)
PE	Manualized; includes in vivo/imaginal exposure, psychoeducation	Strong (first-line)	–	Strong (first-line)
EMDR	8-phase, trauma-focused; uses bilateral stimulation	Strong (first-line)	Strong (first-line in 6/14 guidelines)	Conditional (second-line, due to variable evidence)
NET	Life narrative focusing on traumatic events; suitable for complex trauma	Conditional (second-line)	–	Conditional (second-line)
SSRIs	Pharmacological; includes fluoxetine, paroxetine and sertraline	Conditional (second-line)	Strong (first-line in 13/14 guidelines)	Conditional (second-line)
SNRIs	Pharmacological; includes venlafaxine	Conditional (second-line)	Conditional (venlafaxine first-line in 10/14 guidelines)	Conditional (second-line)
Ketamine	Experimental only[d]	Excluded	Excluded	Excluded (insufficient evidence)
MDMA	Experimental only[d]	Excluded	Excluded	Excluded (insufficient evidence)

CPT cognitive processing therapy, *EMDR* eye movement desensitization and reprocessing therapy, *MDMA* 3,4 – methylenedioxymethamphetamine, *NET* narrative exposure therapy, *PE* prolonged exposure therapy, *SSRIs* selective serotonin reuptake inhibitors, *SNRIs* serotonin–norepinephrine reuptake inhibitors, *TF-CBT* trauma-focused cognitive behaviour therapy

Notes:

[a]NHMRC (2031)

[b]Martin et al. (2021)

[c]APA (2025)

[d]See Table 8.1

Note: Veterans Affairs (United States) guidelines endorse TF-CBT psychotherapies (PE, CPT, CT) and call for hybrid approaches combining psychotherapy with pharmacologic and neuromodulatory techniques

Table 16.3 Food and Drug Administration (FDA) approval status of neurotherapies and pharmacotherapies

Intervention	Category	Approved uses	PTSD uses not approved
Transcranial magnetic stimulation (TMS)	Technology	Several indications, including MDD and OCD[a] fMRI-guided TMS for depression[b]	Neither TMS nor fMRI-guided TMS are approved specifically for PTSD[b]
EEG-guided neurofeedback	Technology	Exception: specific amygdala-EEG neurofeedback device (the prism device by GrayMatters health) for use as an adjunct to psychotherapy and pharmacotherapy for PTSD[c]	Not broadly approved for treating PTSD or other psychiatric conditions[d]
MDMA-assisted therapy	Drug	None	Not approved for PTSD, nor for off-label use[e]
Ketamine	Drug	Anaesthetic[f]	Not approved for any psychiatric disorder, including PTSD[f]
SSRIs	Drug	Sertraline and paroxetine for PTSD and MDD	None
SNRIs	Drug	Several indications, including depression	Not approved for PTSD, although venlafaxine is used off-label

EEG electroencephalography, *fMRI* functional magnetic resonance imaging, *MDD* major depressive disorder, *MDMA* 3,4-methylenedioxymethamphetamine, *OCD* obsessive-compulsive disorder, *PTSD* posttraumatic stress disorder, *SSRIs* selective serotonin reuptake inhibitors, *SNRIs* serotonin–norepinephrine reuptake inhibitors
[a]Cohen et al. (2022)
[b]Gomes et al. (2025)
[c]Zagorsk (2023)
[d]AboutNeurofeedback (n.d.)
[e]Kupferschmidt (2024)
[f]FDA (2023)

16.3 Comparison of the Relative Effectiveness of Neuroscientific, Psychological and Pharmacological Interventions

Beyond effect sizes and recommendations in national guidelines, the clinical descriptions and trial outcomes provide context for selecting treatments. Table 16.4 summarizes the key interventions considered in this work, describing the mechanism, effectiveness and longevity of treatment effect for each intervention.

The evidence converges on the superiority of trauma-focused psychotherapies (TF-CBTs) over pharmacotherapies in achieving larger effect sizes and higher rates

Table 16.4 Comparison of key psychotherapies, pharmacotherapies and neurotherapies: mechanism, effectiveness and longevity of treatment

Intervention	Type	Mechanism	Effectiveness	Longevity of effect
fMRI-guided TMS[a]	Neurotherapy	Targets brain regions identified via fMRI (e.g. dlPFC and amygdala circuit)	High effectiveness when precisely targeted; improved outcomes over unguided TMS	Long-lasting effects reported; may require periodic boosters
EEG-based neurofeedback[b]	Neurotherapy	Self-regulation of brain activity through real-time EEG feedback	Approximately 70–80% of patients no longer meet PTSD criteria after training	May last several months to years; some patients need periodic retraining
EMDR therapy[c]	Psychotherapy	Uses bilateral stimulation (e.g. following the therapist's fingers while recalling traumatic event)	Highly effective; up to 90% reduction of all symptoms after a single trauma	Sustained effects with follow-up; booster sessions sometimes required
PE therapy[d]	Psychotherapy	Repeatedly confronts trauma-related memories and situations	Effective; up to 85% in reducing most PTSD symptoms	Can last months to years if therapy is completed
CBT/CPT[e]	Psychotherapy	Restructures negative thought patterns	50–64% decline in those with PTSD	Moderate retention; often needs reinforcement
Cognitive therapy[f]	Psychotherapy	Targets maladaptive beliefs and cognitive distortions without emphasis on exposure	Effective in reducing PTSD symptoms, especially cognitive symptoms (approx. 60–70% improvement)	Effects maintained with booster sessions; less effective if traumatic memory remains unprocessed
MDMA[g]	Pharmacotherapy	Facilitates emotional openness during therapy	70–80% no longer meet criteria for PTSD	Long-term effects possible when used with psychotherapy
Ketamine[h]	Pharmacotherapy	NMDA receptor antagonist	Provides rapid relief of about 50% in PTSD symptoms	Short-lived (from days up to 2 weeks); repeated infusions required
SSRIs[i]	Pharmacotherapy	Increases serotonin levels to stabilize mood	Moderately effective; 50–70% improvement in PTSD symptoms	Variable; often needs maintenance therapy

(continued)

Table 16.4 (continued)

Intervention	Type	Mechanism	Effectiveness	Longevity of effect
SNRIs[j]	Pharmacotherapy	Affects serotonin and norepinephrine levels	Less than 50% improve their PTSD symptoms	Limited long-term benefit; many relapse

CBT cognitive behaviour therapy, *CPT* cognitive processing therapy, *dlPFC* dorsolateral prefrontal cortex, *EEG* electroencephalography, *EMDR* eye movement desensitization and reprocessing, *fMRI* functional magnetic resonance imaging, *MDMA* 3,4-methylenedioxymethamphetamine, *PE* prolonged exposure, *PTSD* posttraumatic stress disorder, *SNRIs* serotonin–norepinephrine reuptake inhibitors, *SSRIs* selective serotonin reuptake inhibitors, *TMS* transcranial magnetic stimulation

Notes:

[a]See, e.g. Fox et al. (2012) and Siddiqi et al. (2020)

[b]See, e.g. van der Kolk et al. (2016) and Peniston and Kulkosky (1991)

[c]See, e.g. Shapiro (2001), van der Kolk et al. (2007) and Bisson et al. (2015)

[d]See, e.g. Foa (2000) and Powers et al. (2010)

[e]See, e.g. Resick et al. (2002), Monson et al. (2006) and Bisson et al. (2015)

[f]See, e.g. Ehlers et al. (2005) and Clark and Ehlers (2004)

[g]See, e.g. Mithoefer et al. (2011)

[h]See, e.g. Feder et al. (2014)

[i]See, e.g. Cukor et al. (2010)

[j]See, e.g. Davidson (2006)

of clinical remission. This supports guidelines that designate TF-CBT (PE, CPT, CT) and EMDR as first-line treatments for PTSD and agrees with the conclusions presented in Sects. 7.2 and 7.3.

Pharmacotherapies retain an important role, especially for patients unable to access or tolerate psychotherapy. SSRIs and SNRIs remain the most studied, with moderate efficacy, while the novel agents ketamine and MDMA offer rapid and substantial symptom relief in treatment-resistant cases.[5] MDMA-assisted psychotherapy, in particular, yields remission rates up to 80%, suggesting a paradigm shift towards pharmacologically augmented psychotherapy.

Transcranial magnetic stimulation (TMS) and EEG-guided neurofeedback, despite their 'experimental' status, also show high and long-lasting effectiveness, especially TMS. This suggests the combination of TMS, neurofeedback and ketamine, or MDMA could offer a significant improvement for PTSD treatment for select individual patients.

[5] Feder et al. (2014) and Mithoefer et al. (2011).

16.4 Conclusion

EMDR therapy and trauma-focused psychotherapies have the most important effects in reducing PTSD, with the strongest effects among the latter being TF-CBT, CT and PE therapies.[6]

Trauma-focused psychotherapies are generally recommended as first-line interventions, with the most support for cognitive- and exposure-based approaches, such as PE, CT or CPT,[7] as well as EMDR.[8] There is general agreement that these psychotherapeutic interventions are substantially more effective than pharmacological interventions in reducing PTSD.[9] The important exceptions to this claim are interventions with ketamine and MDMA, but these are not yet FDA approved or recommended for PTSD.

References

AboutNeurofeedback (n.d.) FDA requirements for neurofeedback software and equipment. https://www.aboutneurofeedback.com/neurofeedback-info-center/faq/fda-requirements-for-neurofeedback-software-and-equipment/

APA (American Psychological Association) (2025) Clinical practice guideline for the treatment of posttraumatic stress disorder (PTSD) in adults. APA, Washington, DC. https://www.apa.org/ptsd-guideline/ptsd.pdf

Bisson JI, Cosgrove S, Lewis C, Robert NP (2015) Post-traumatic stress disorder. BMJ 351:h6161. https://doi.org/10.1136/bmj.h6161

Charney ME, Hellberg SN, Bui E, Simon NM (2018) Evidenced-based treatment of posttraumatic stress disorder: an updated review of validated psychotherapeutic and pharmacological approaches. Harv Rev Psychiatry 26(3):99–115. https://doi.org/10.1097/hrp.0000000000000186

Chen YR, Hung KW, Tsai JC, Chu H, Chung MH, Chen SR, Liao YM, Ou KL, Chang YC, Chou KR (2014) Efficacy of eye-movement desensitization and reprocessing for patients with posttraumatic-stress disorder: a meta-analysis of randomized controlled trials. PLoS One 9(8):e103676. https://doi.org/10.1371/journal.pone.0103676

Clark DM, Ehlers A (2004) Posttraumatic stress disorder: from cognitive theory to therapy. In: Leahy RL (ed) Contemporary cognitive therapy: theory, research, and practice. Guilford Press, New York, pp 141–160

Cohen S, Bikson M, Badran B, George M (2022) A visual and narrative timeline of US FDA milestones for transcranial magnetic stimulation (TMS) devices. Brain Stim 15(1):73–75. https://doi.org/10.1016/j.brs.2021.11.010

Cukor J, Olden M, Lee F, Difede J (2010) Evidence-based treatments for PTSD, new directions, and special challenges. Ann N Y Acad Sci. 1208:82–89. https://doi.org/10.1111/j.1749-6632.2010.05793

[6] Lewis et al. (2020).

[7] Charney et al. (2018).

[8] Chen et al. (2014).

[9] Merz et al. (2019) and Charney et al. (2018).

Davidson JR (2006) Pharmacologic treatment of acute and chronic stress following trauma: 2006. J Clin Psychiatry 67(Suppl 2):34–39

Ehlers A, Clark DM, Hackmann A, McManus F, Fennell M (2005) Cognitive therapy for posttraumatic stress disorder: development and evaluation. Behav Res Ther 43(4):413–431. https://doi.org/10.1016/j.brat.2004.03.006

FDA (US Food and Drug Administration) (2023) FDA warns patients and health care providers about potential risks associated with compounded ketamine products, including oral formulations, for the treatment of psychiatric disorders. 10 Oct. https://www.fda.gov/drugs/human-drug-compounding/fda-warns-patients-and-health-care-providers-about-potential-risks-associated-compounded-ketamine

Feder A, Parides MK, Murrough JW, Perez AM, Morgan JE, Saxena S, Kirkwood K, Aan Het Rot M, Lapidus KA, Wan LB, Iosifescu D, Charney DS (2014) Efficacy of intravenous ketamine for treatment of chronic posttraumatic stress disorder: a randomized clinical trial. JAMA Psychiatry 71(6):681–688. https://doi.org/10.1001/jamapsychiatry.2014.62

Foa EB (2000) Psychosocial treatment of posttraumatic stress disorder. J Clin Psychiatry 61(Suppl 5):43–48. discussion 49–51

Fox MD, Buckner RL, White MP, Greicius MD, Pascual-Leone A (2012) Efficacy of transcranial magnetic stimulation targets for depression is related to intrinsic functional connectivity with the subgenual cingulate. Biol Psychiatry 72(7):595–603. https://doi.org/10.1016/j.biopsych.2012.04.028

Gomes P, Imtiaz A, Groden M, Zaidi M (2025) A review and comparison of FDA-approved transcranial magnetic stimulation (TMS) protocols for depression. CNS Spectr 29(5):507–508. https://doi.org/10.1017/S1092852924001767

Kupferschmidt K (2024) FDA rejected MDMA-assisted PTSD therapy. Other psychedelics firms intend to avoid that fate. ScienceInsider, 12 Aug. https://www.science.org/content/article/fda-rejected-mdma-assisted-ptsd-therapy-other-psychedelics-firms-intend-avoid-fate

Lewis C, Roberts NP, Andrew M, Starling E, Bisson JI (2020) Psychological therapies for posttraumatic stress disorder in adults: systematic review and meta-analysis. Eur J Psychotraumatol 11(1):1729633. https://doi.org/10.1080/20008198.2020.1729633

Martin A, Naunton M, Kosari S, Peterson G, Thomas J, Christenson JK (2021) Treatment guidelines for PTSD: a systematic review. J Clin Med 10(18):4175. https://doi.org/10.3390/jcm10184175

Merz J, Schwarzer G, Gerger H (2019) Comparative efficacy and acceptability of pharmacological, psychotherapeutic, and combination treatments in adults with posttraumatic stress disorder: a network meta-analysis. JAMA Psychiatry 6(9):904–913. https://doi.org/10.1001/jamapsychiatry.2019.0951

Mithoefer MC, Wagner MT, Mithoefer AT, Jerome L, Doblin R (2011) The safety and efficacy of {+/−}3,4−methylenedioxymethamphetamine-assisted psychotherapy in subjects with chronic, treatment-resistant posttraumatic stress disorder: the first randomized controlled pilot study. J Psychopharmacol 25(4):439–452. https://doi.org/10.1177/0269881110378371

Monson CM, Schnurr PP, Resick PA, Friedman MJ, Young-Xu Y, Stevens SP (2006) Cognitive processing therapy for veterans with military-related posttraumatic stress disorder. J Consult Clin Psychol 74(5):898–907. https://doi.org/10.1037/0022-006x.74.5.898

NHMRC (National Health and Medical Research Council) (2031) Interventions. In: Australian guidelines for the prevention and treatment of acute stress disorder, posttraumatic stress disorder and complex PTSD. Phoenix Australia, ch 4. https://www.phoenixaustralia.org/australian-guidelines-for-ptsd/

Peniston EG, Kulkosky PJ (1991) Alpha-theta brainwave neurofeedback therapy for Vietnam veterans with combat-related posttraumatic stress disorder. Med Psychother 4:47–60

Powers MB, Halpern JM, Ferenschak MP, Gillihan SJ, Foa EB (2010) A meta-analytic review of prolonged exposure for posttraumatic stress disorder. Clin Psychol Rev 30(6):635–641. https://doi.org/10.1016/j.cpr.2010.04.007

Resick PA, Nishith P, Weaver TL, Astin MC, Feuer CA (2002) A comparison of cognitive-processing therapy with prolonged exposure and a waiting condition for the treatment of chronic posttrau-

matic stress disorder in female rape victims. J Consult Clin Psychol 70(4):867–879. https://doi.org/10.1037/0022-006x.70.4.867
Shapiro F (2001) Eye movement desensitization and reprocessing: basic principles, protocols, and procedures, 2nd edn. Guilford Press, New York
Siddiqi SH, Taylor SF, Cooke D, Pascual-Leone A, George MS, Fox MD (2020) Distinct symptom-specific treatment targets for circuit-based neuromodulation. Am J Psychiatry 177(5):435–446. https://doi.org/10.1176/appi.ajp.2019.19090915
van der Kolk BA, Spinazzola J, Blaustein ME, Hopper JW, Hopper EK, Korn DL, Simpson WB (2007) A randomized clinical trial of eye movement desensitization and reprocessing (EMDR), fluoxetine, and pill placebo in the treatment of posttraumatic stress disorder: treatment effects and long-term maintenance. J Clin Psychiatry 68(1):37–46. https://doi.org/10.4088/jcp.v68n0105
van der Kolk BA, Hodgdon H, Gapen M, Musicaro R, Suvak MK, Hamlin E, Spinazzola J (2016) A randomized controlled study of neurofeedback for chronic PTSD. PLoS One 11(12):e0166752. https://doi.org/10.1371/journal.pone.0166752
Van Etten ML, Taylor S (1998) Comparative efficacy of treatments for post-traumatic stress disorder: a meta-analysis. Clin Psychol Psychother 5:126–144
Watts BV, Schnurr PP, Mayo L, Young-Xu Y, Weeks WB, Friedman MJ (2013) Meta-analysis of the efficacy of treatments for posttraumatic stress disorder. J Clin Psychiatry 74(6):e541–e550. https://doi.org/10.4088/jcp.12r08225
Yunitri N, Chu H, Kang XL, Wiratama BS, Lee TY, Chang LF, Liu D, Kustanti CY, Chiang KJ, Chen R, Tseng P, Chou KR (2023) Comparative effectiveness of psychotherapies in adults with posttraumatic stress disorder: a network meta-analysis of randomised controlled trials. Psychol Med 53(13):6376–6388. https://doi.org/10.1017/s0033291722003737
Zagorsk N (2023) FDA clears neurofeedback intervention for PTSD. Psychiatr News 58(8). https://doi.org/10.1176/appi.pn.2023.08.8.60

Part VII
The Synaptic Origins of PTSD

Chapter 17
Dissociative PTSD Explained by Failure of Dendritic Integration of Synapses of Cortical Layer 5 Pyramidal Neurons

17.1 The Central Role of Cortical Layer 5 Pyramidal Neurons (L5PNs) in Consciousness

Thalamic neurons classified as specific, as they receive particular sensory inputs, mostly project to the basal dendrites of cortical layer 5 pyramidal neurons (L5PNs) (Figs. 17.1 and 17.2b).[1] In contrast, thalamic neurons classified as non-specific, as they receive diffuse cortical inputs, project diffusely to the apical tufts of L5PNs (Fig. 17.2b), forming together a thalamocortical module (Fig. 17.2b). The cortex may be considered as a collection of such modules (Fig. 17.2c), each making an associative unit. The association comes about from the non-specific thalamic input delivering cortical information concerning context to the apical tuft dendrites of L5PNs and the sensory input to their basal dendrites; calcium-dependent activity in the apical tuft can spread in a back-propagation to the pyramidal cell body where it, together with the sensory input-initiated action potentials, greatly enhances the spiking activity and the output of the L5PNs to subcortical sites. The apical tufts, besides receiving the non-specific input from thalamic nuclei, also receive cortico-cortical inputs from higher cortical areas subserving contextual information and attention as well as neuromodulatory inputs from subcortical areas (such as the dorsal raphe nucleus and the locus coeruleus). These properties of L5PNs are thought to provide a mechanism by which sensory (perceived) experiences, received through the basal dendrites, are associated with an existing 'internal representation' of the world, received through the apical dendrites.[2]

Depolarizing the apical tuft dendrites of L5PNs sometimes causes enhanced action potential activity in the cell body, which can be blocked by general anaesthetics as well as by metabotropic glutamate receptor and cholinergic receptor

[1] Llinás et al. (1998) and Ledergerber and Larkum (2010).

[2] Larkum (2013).

M. Bennett, *Veteran's Stress, Trauma and Synapse Rehabilitation*,
https://doi.org/10.1007/978-981-95-9594-5_17

a

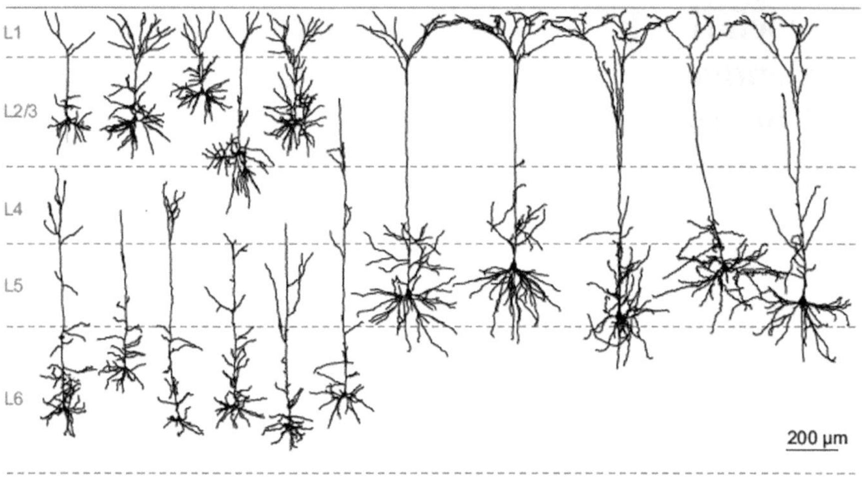

b

Pyramidal cell type	Layer 6	Layer 2/3	Layer 5
Average length apical tree (μm)	963 ± 160	482 ± 37	1341 ± 154
Average diameter apical tree (μm)	1.33 ± 0.22	1.57 ± 0.46	3.35 ± 0.61
Average length apical shaft (μm)	772 ± 111	323 ± 244	739 ± 186
Average diameter apical shaft (μm)	1.49 ± 0.33	2.11 ± 1.02	4.94 ± 1.10
Effective length constant soma → dend (shaft)	501	556 †	438 ‡
Effective electrotonic length soma → dend (shaft)	1.54	0.58	1.69
Effective length constant dend (shaft) → soma	334	302 †	457 ‡
Effective electrotonic length dend (shaft) → soma	2.31	1.07	1.62

Fig. 17.1 Anatomical comparison of apical dendrites of different pyramidal neurons in the neocortex showing (**a**) neurolucida reconstructions of pyramidal neurons from layers 2/3, 5 and 6 of the somatosensory cortex in rats and (**b**) table of the average data for five neurons from L2/3, L5 and L6 neurons from L6 shown in **a**. (Reproduced from Ledergerber and Larkum (2010), with permission from The Society for Neuroscience. Creative Commons CC BY 4.0 (https://creativecommons.org/licenses/by/4.0/))

antagonists. These effectively block the enhanced spiking at the cell body due to depolarization of the tufts. This led to the concept that a ‘gate’ must exist along L5PN apical shafts proximal to the dendritic tufts. Such a theory was tested in the somatosensory cortex in which higher-order thalamic nuclei, or posteromedial (POm) nuclei, project to the shaft, where they can gate the (enhanced) influence of the tuft dendrites on the cell body (soma), thereby leading to an increase in the output of L5PNs. If the influence of Pom nuclei in enhancing the efficacy of tuft dendrites on soma spiking is blocked by a general anaesthetic, then L5PN burst spiking is blocked, leading to a decrease in the output of the cortical column (Fig. 17.3). Given that the dendritic tufts receive a depolarizing synaptic input from a wide range of cortex, the net result is decoupling of activity over a widespread region of cortex and thus unconsciousness.

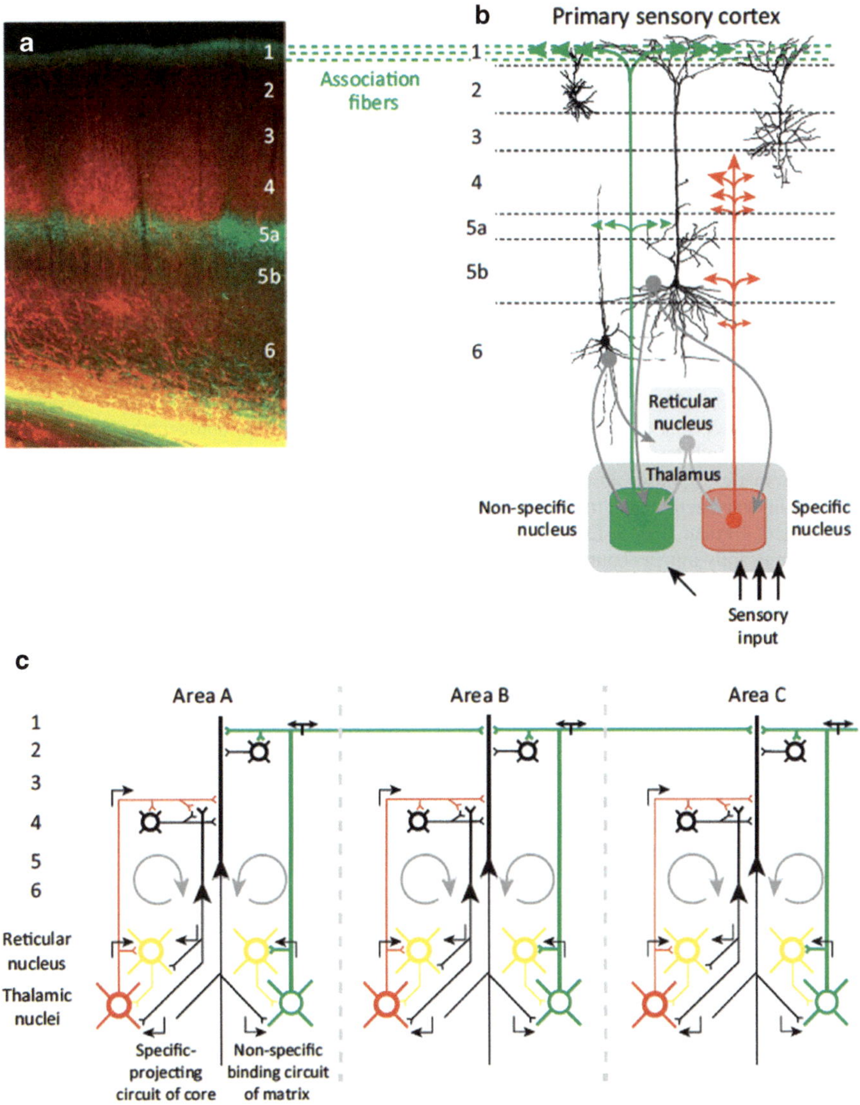

Fig. 17.2 (**a**) Axons from the primary, specific (ventral posterior medial, red) and secondary, non-specific (posterior medial, green) thalamic nuclei, visualized via injection of an adeno-associated virus (AAV) conjugated with green and red fluorescent proteins to the respective thalamic regions. (**b**) Schematic representation of the long-range input to the primary sensory cortex in rats. The colour scheme follows the fluorescent protein colours in (**a**) for easy comparison. Association fibres carrying feedback information from other cortical areas are shown overlapping with thalamic, non-specific input in layer 1. (**c**) A schematic representation of a chain of cortical modules. 'Core', specific-projecting neurons (red), and 'matrix', non-specific binding circuit neurons (green), interact with the cortex in different lamina via the dendrites of pyramidal and local inhibitory neurons (black). Inhibitory neurons of the reticular nucleus (yellow) control the mode of firing thalamic cortico-projecting neurons promoting oscillatory behavior (arrows). This architecture forms repeating modules synchronizing cell assemblies across the cortex and thalamus. (Reproduced from Larkum (2013) with permission from Elsevier)

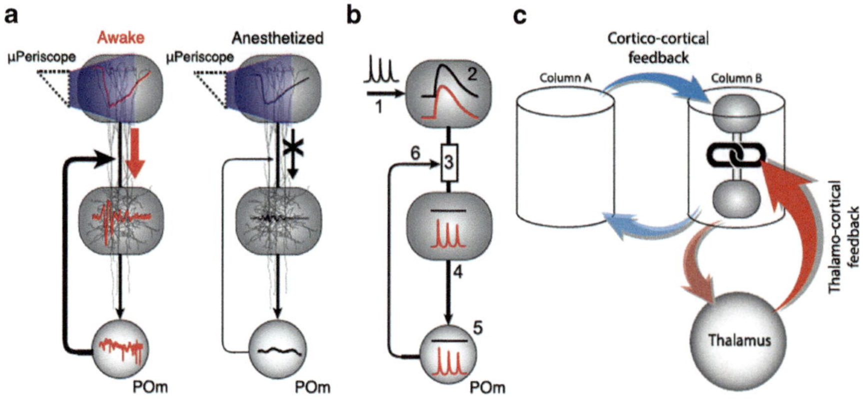

Fig. 17.3 A theory of general anaesthesia showing the following: (**a**) optogenetic stimulation of distal apical dendrites resulted in depolarization of dendrites in both anaesthetized and awake states. Somatic responses, however, clearly depended on the conscious state. Elevated neuronal activity in the posteromedial thalamic nucleus (Pom) during wakefulness correlates with enhanced coupling of cortical layer 5 (L5) pyramidal neurons, and inactivation of the POm during wakefulness significantly impairs coupling. (**b**) An interpretation of the present findings. Feedback signals coming to distal apical dendrites (1) cause equally strong excitatory postsynaptic potentials (2). Because the marked difference in dendro-somatic coupling (3), somatic action potentials are reliably evoked during wakefulness, whereas delayed and few, if any, action potentials are evoked during anaesthesia (4). Higher POm activity (5) enhances coupling of L5 pyramidal neurons during wakefulness (6). (**c**) The model that reconciles the two long-standing hypotheses regarding anaesthetic-induced unconsciousness. One hypothesis posits that the cortico-cortical feedback loop is crucial for conscious perception; the other emphasizes the importance of the thalamocortical loop. Recent observations reveal that decoupling of dendro-somatic compartments of L5 pyramidal neurons results in breakdown of both cortico-cortical and thalamocortical feedback loops, providing a cellular mechanism that reconciles the two apparently distinct hypotheses. (Reproduced from Suzuki and Larkum (2020) with permission from Elsevier)

The apical dendritic tufts of L5PNs receive synaptic inputs from both the higher-order (non-specific) thalamus and higher cortical areas (together with modulating inputs), and the basal dendrites of the L5PNs receive specific feedforward synaptic inputs from lower cortical areas or, in the case of sensory cortical areas, like the visual cortex and the somatosensory cortex, from thalamocortical relay cells. If both apical tuft and basal dendrites are depolarized at the same time, there is often significant enhancement of the spike firing at the cell body that occurs in bursts. A non-specific thalamic neuron synapses on many L5PN apical tufts so that the non-specific thalamus can evoke activity, through its action on L5PNs, across an area of the cortex.[3]

If an animal is awake, in a state of intransitive consciousness but tranquil, there is an approximately fourfold increase in the activity of the apical tuft synapses, compared with when the animal is anaesthetized. In contrast, if the animal responds to a signal (like lifting of a limb in response to a puff of air on the limb), there is a

[3] See Aru et al. (2019), Fig. 1.

14-fold increase in the response that lasts for seconds, which does not happen if the air puff is subliminal.[4] This is interpreted as the L5PN apical tufts giving only a fourfold increase in calcium activity (associated with intransitive consciousness), but when actually perceiving the puff, a 14-fold tuft response is involved (transitive consciousness).[5] Thus, the state and contents of consciousness (i.e. intransitive and transitive consciousness, with the latter dependent on the former[6]) are both dependent on L5PNs, which in turn are central to both cortico-cortical and thalamocortical loops.[7] As for the condition of unconscious brain activity, such as motor activity guiding a person's writing hand, this involves cortical areas that do not possess L5PNs, such as the cerebellum and basal ganglia. Indeed, cortical processing per se that does not involve L5PN and non-specific thalamic neurons (the L5PN/non-specific thalamic loop) does not give rise to consciousness.

In summary, L5PNs are common to both cortico-cortical and thalamocortical loops, the former for intransitive consciousness and the latter for transitive consciousness, binding the two together, with transitive consciousness dependent on intransitive consciousness.

17.2 Models of the Role of L5PNs in Consciousness

17.2.1 Dendritic Integration Theory

The attractiveness of the dendritic integration theory (DIT) proposed by Larkum and colleagues lies in its combination of systems-level ideas with specific mechanisms concerning the biology of single neurons, specifically pyramidal neurons with their cell bodies in cortical layer 5 (L5PN). These are called thick-tufted neurons because of their prolific apical dendrites found in layer 1 atop a long apical trunk, which electrotonically separates the tuft dendrites from the basal dendrites near the cell body (soma) in layer 5 (see Fig. 17.1). The apical tuft dendrites possess synapses principally from other cortical sites (40–50% of apical tuft synapses) and from non-specific thalamic neurons (15–25% of apical tuft synapses) as well as modulatory neurons (20–30% of apical tuft synapses). In contrast, specific projecting thalamic neurons subserving specific sensory information mainly synapse on layer 4/5 basal dendrites of L5PNs.[8] Synapses formed by the non-specific thalamus spread diffusely throughout extended areas of layer 1, unlike the specific thalamic input to precise sites in layers 4/5. Conceptually, the input to the dendritic tuft receives cortical information concerning context, whereas the basal dendrites

[4] Aru et al. (2019).

[5] Aru et al. (2019).

[6] Bennett and Hacker (2022).

[7] Aru et al. (2019).

[8] See Munn et al. (2023), Fig. 1.

receive input concerning, in many cases, sensory experience. Of considerable importance for cortical functioning is the recent discovery of oblique dendrites that emanate partway along the apical trunk, closer to the basal somatic dendrites than to the apical tufts. A principal synaptic input to these dendrites is collaterals of the non-specific thalamic input.

L5PNs form collateral synapses with other L5PNs in the same layer, and apical tuft dendrites of different L5PNs receive common non-specific thalamic (e.g. intralaminar nuclei) and cortico-cortical feedback inputs, all of which synchronized activity across L5PNs. The resulting spatial correlation of apical inputs across L5PNs (high sigma) promotes synchronized dendritic calcium spikes, thus triggering burst firing at the soma, conditional on the oblique dendrites facilitating these spikes reaching the basal region through input onto them from the thalamus, locus coeruleus and basal forebrain (nucleus of Meynert).

L5PNs provide the main projections out of the prefrontal cortex (PFC), primarily to subcortical areas.[9] In the case of medial PFC (mPFC) L5PNs, these are 27% to the lateral hypothalamus, 23% to the ventral striatum, 18% to the dorsal striatum and 8% to the basolateral amygdala (also projections to the thalamus, superior colliculus, pons, brain stem and spinal cord). In particular, higher-order thalamic neurons are targeted, making these neurons strategically placed with their large spreading apical dendrites to integrate L5PNs (Fig. 17.2).[10]

Synaptic activity on the basal dendrites of L5PNs (principally driven by specific thalamic input) gives rise to regular action potential (spike) firing, but intense synaptic activity to the tuft dendrites (facilitated by NMDA spikes and calcium plateau action potentials) is necessary to overcome the electrotonic distance to the soma. If there are coincident tuft and basal dendrite inputs, then high-frequency burst action potential firing at the soma occurs. However, this is conditional on the oblique dendrites being activated by their synapses (with metabotropic glutamate receptors), principally from the non-specific thalamic neurons, which provides a facilitatory mechanism that effectively shortens the electrotonic distance, so that blocking this oblique dendrite activity (e.g. with anaesthetics) leads to failure of coincident tuft and basal dendritic inputs to generate burst action potential firing at the soma that acts as if it only receives input from the basal dendrites.

A biophysical model of the above structure and function was developed, in which model L5PNs give asynchronous regular action potential firing when stimulated via basal dendrites; this changed to burst firing of action potentials when there was coincident apical tuft dendritic activity, conditional on the facilitation provided by the oblique dendrites enhancing the reach of the tuft dendritic activity to the soma (a function principally of the non-specific thalamic input to the oblique dendrites).[11] In the model, this function is given the value beta, with beta = 0 when all soma action potentials are single and of low frequency, and beta = 1 when all action

[9] Ding et al. (2001).

[10] Gabbott et al. (2005).

[11] Munn et al. (2023).

potentials are in high-frequency bursts. The non-specific thalamic projections to the oblique dendrites also provide spatially diffuse inputs to the tuft apical dendrites of many L5PNs. To allow for this spatial and temporal coupling across L5PNs, mediated by the non-specific thalamic cortical neurons, which has the potential to produce transient neuronal ensembles of L5PNs, the factor sigma was introduced, which provides a measure of this coupling; when sigma is a maximum (= N), all the L5PN neurons are synchronously coupled within the ensemble and driven by the non-specific thalamic neurons and the L5PNs give synchronous bursts. In contrast, when sigma = 0, the L5PNs are not synchronously coupled. This model gives synchronous bursts in these transient ensembles.

17.2.2 Integrated Information Theory Applied to Dendritic Integration Theory

According to Tononi's integrated information theory (IIT),[12] intransitive consciousness can be objectively measured in a system by determining the value of a factor phi. Large phi indicates that the system has more integrated information than the sum of its separated parts.[13] Applying IIT to the DIT model shows that at relatively low spatio-temporal coupling of L5PNs (i.e. low sigma), then phi increases with increasing beta (i.e. increased efficiency of the tuft input effects on the soma).[14] However, if there is relatively high spatio-temporal coupling (i.e. high sigma, comparable to N), a large monotonic increase in phi occurs with increasing beta reaching a maximum at beta = 0.3, and there is synchronous burst firing in the network. If phi is related to intransitive consciousness, then the DIT model can produce large phi and thus intransitive consciousness.

The DIT offers a solution to the question as to why the 'contents of consciousness' (i.e. transitive consciousness) are closely linked to the 'state of consciousness' (i.e. intransitive consciousness).[15] Intransitive consciousness depends on cortico-cortical loops maintained by the non-specific thalamus, which receives cortical input and maintains extensive excitation of L5PNs apical dendrites through its own thalamic-L5PN loop. Non-specific thalamic input to apical dendrites and to the oblique dendrites of L5PNs is necessary to raise the depolarization of their soma to a level that allows the specific thalamic input that carries sensory information to give burst action potential firing at a rate that allows for transitive (object) consciousness. Both intransitive and transitive consciousness are linked through their common dependence on L5PNs, with its three specialized compartments of apical tuft, oblique and basal dendrites.

---

[12] Tononi et al. (2016).

[13] See Haun et al. (2017), Fig. 2.

[14] Munn et al. (2023), Fig. 2.

[15] Aru et al. (2019).

17.3 Effects of Stress and Trauma on L5PNs

It is well established that L5PNs retract apical dendrites and synaptic spines when subjected to chronic restraint stress (mPFC, infralimbic and prelimbic mPFC[16]) as well as acute stress (barrel cortex[17]). The stress signalling mechanism that leads to the loss of spines involves the action of brain-derived neurotrophic factor (BDNF) and mTORC.[18] Chronic stress reduces dendritic length and spine density in apical tufts,[19] with little change in the length of the apical shaft, at least in L2/3 neurons.[20] Single prolonged stress of rodents gives a long-lasting reduction in the excitatory synaptic drive onto L5PN apical dendrites of the infralimbic cortex (ventromedial PFC).[21] This is very likely due to the loss of synaptic spines along the dendrites of L5PNs,[22] consequent on the elevation of glucocorticoids.[23]

Studies on the effects of stress on layer 2/3 pyramidal neurons in rodent mPFC indicate a significant reduction, on the order of 20%, in both the length and number of branches of apical dendrites in layer I, but no changes in basal dendrites.[24] Of great interest is the recent discovery that the stress-induced loss of synapses in particular cortical layers is accompanied by complement microglia activation that is restricted to these layers, suggesting that the latter gives rise to the former.[25] This idea is supported by the observation that rodents deprived of complement component C3 show reduction in activated microglia and synapse loss in specific layers (2/3).[26]

17.4 The Effect of Loss of L5PN Dendritic Synapses on Consciousness and PTSD

The coincident depolarization of apical tuft-like dendrites of L5PN with that of basal dendrites can trigger calcium action potentials that facilitate somatic action potential bursting, if the oblique dendrites are also activated (the 'gate') and beta is

[16] Goldwater et al. (2009); Perez-Cruz et al. (2007); Liu and Aghajanian (2008); Ramkumar et al. (2012); Izquierdo et al. (2006)

[17] Chen et al. (2018); but see Urban et al. (2019) for increased apical dendritic length in the mPFC of male rodents following social stress.

[18] Woo et al. (2021).

[19] McEwen and Morrison (2013).

[20] Shansky and Morrison (2009).

[21] Nawreen et al. (2021).

[22] Chen et al. (2018).

[23] Madalena and Lerch (2017) and Anderson et al. (2016).

[24] Radley et al. (2004), Liston et al. (2006), and Cook and Wellman (2004).

[25] Tillmon et al. (2024).

[26] Tillmon et al. (2024).

increased from zero maximizing phi and amplifying the effects on the soma of sensory stimuli to the basal dendrites, thereby allowing the sensory stimuli to reach conscious awareness (transient consciousness). With a decrease in apical dendrites as a consequence of stress, depolarizations necessary to achieve calcium plateau potentials are less likely, thereby the burst firing at the soma is less likely, giving rise to regular spiking (action potentials) with the consequences that sensory stimuli lose their impact on L5PNs and fail to be integrated with the cortico-cortical inputs to the apical dendrites and hence lose context.

In summary, a loss of apical dendrites and their synapses leads to a decrease in depolarization at the apical tuft dendrites and therefore of the probability of action potential burst firing at the soma in response to basal dendritic activity as a consequence of sensory inputs. Such sensory inputs must now increase significantly in order for the soma to produce action potential bursting necessary for perceptual awareness. Concomitant with this is a decrease in phi (intransitive consciousness). This explains dissociative posttraumatic stress disorder (PTSD) in which normal sensory inputs fail to trigger conscious perception, although intransitive consciousness (phi) is still present but also lowered. It explains why in dissociative PTSD the world feels unreal due to weak processing of the sensory input to the basal dendrites (derealization), reduced response to stimuli (decrease in awareness) and failure of L5PNs to excite subcortical areas (and so emotional detachment). It also helps explain why there is a strong correlation ($r = 0.52$) between current PTSD severity and the emergence of 'depersonalization' and 'derealization'.[27]

References

Anderson RM, Glanz RM, Johnson SB, Miller MM, Romig-Martin SA, Radley JJ (2016) Prolonged corticosterone exposure induces dendritic spine remodeling and attrition in the rat medial prefrontal cortex. J Comp Neurol 524(18):3729–3746. https://doi.org/10.1002/cne.24027

Aru J, Suzuki M, Rutiku R, Larkum ME, Bachmann T (2019) Coupling the state and contents of consciousness. Front Syst Neurosci 13:43. https://doi.org/10.3389/fnsys.2019.00043

Bennett MR, Hacker PMS (2022) Transitive and intransitive consciousness. In: Philosophical foundations of neuroscience, 2nd edn. Wiley Blackwell, Hoboken, pp 265–291

Chen CC, Lu J, Yang R, Ding JB, Zuo Y (2018) Selective activation of parvalbumin interneurons prevents stress-induced synapse loss and perceptual defects. Mol Psychiatry 23(7):1614–1625. https://doi.org/10.1038/mp.2017.159

Cook SC, Wellman CL (2004) Chronic stress alters dendritic morphology in rat medial prefrontal cortex. J Neurobiol 60(2):236–248. https://doi.org/10.1002/neu.20025

Ding DC, Gabbott PL, Totterdell S (2001) Differences in the laminar origin of projections from the medial prefrontal cortex to the nucleus accumbens shell and core regions in the rat. Brain Res 917(1):81–89. https://doi.org/10.1016/s0006-8993(01)02912-2

Gabbott PL, Warner TA, Jays PR, Salway P, Busby SJ (2005) Prefrontal cortex in the rat: projections to subcortical autonomic, motor, and limbic centers. J Comp Neurol 492(2):145–177. https://doi.org/10.1002/cne.20738

[27] Powers et al. (2016).

Goldwater DS, Pavlides C, Hunter RG, Bloss EB, Hof PR, McEwen BS, Morrison JH (2009) Structural and functional alterations to rat medial prefrontal cortex following chronic restraint stress and recovery. Neuroscience 164(2):798–808. https://doi.org/10.1016/j.neuroscience.2009.08.053

Haun AM, Oizumi M, Kovach CK, Kawasaki H, Oya H, Howard MA, Adolphs R, Tsuchiya N (2017) Conscious perception as integrated information patterns in human Electrocorticography. eNeuro 4(5):ENEURO.0085-17.2017. https://doi.org/10.1523/eneuro.0085-17.2017

Izquierdo A, Wellman CL, Holmes A (2006) Brief uncontrollable stress causes dendritic retraction in infralimbic cortex and resistance to fear extinction in mice. J Neurosci 26(21):5733–5738. https://doi.org/10.1523/jneurosci.0474-06.2006

Larkum M (2013) A cellular mechanism for cortical associations: an organizing principle for the cerebral cortex. Trends Neurosci 36(3):141–151. https://doi.org/10.1016/j.tins.2012.11.006

Ledergerber D, Larkum ME (2010) Properties of layer 6 pyramidal neuron apical dendrites. J Neurosci 30(39):13031–13044. https://doi.org/10.1523/jneurosci.2254-10.2010

Liston C, Miller MM, Goldwater DS, Radley JJ, Rocher AB, Hof PR, Morrison JH, McEwen BS (2006) Stress-induced alterations in prefrontal cortical dendritic morphology predict selective impairments in perceptual attentional set-shifting. J Neurosci 26(30):7870–7874. https://doi.org/10.1523/jneurosci.1184-06.2006

Liu RJ, Aghajanian GK (2008) Stress blunts serotonin- and hypocretin-evoked EPSCs in prefrontal cortex: role of corticosterone-mediated apical dendritic atrophy. Proc Natl Acad Sci U S A. 105(1):359–364. https://doi.org/10.1073/pnas.0706679105

Llinás R, Ribary U, Contreras D, Pedroarena C (1998) The neuronal basis for consciousness. Philos Trans R Soc Lond Ser B Biol Sci 353(1377):1841–1849. https://doi.org/10.1098/rstb.1998.0336

Madalena KM, Lerch JK (2017) The effect of glucocorticoid and glucocorticoid receptor interactions on brain, spinal cord, and glial cell plasticity. Neural Plast 2017:8640970. https://doi.org/10.1155/2017/8640970

McEwen BS, Morrison JH (2013) The brain on stress: vulnerability and plasticity of the prefrontal cortex over the life course. Neuron 79(1):16–29. https://doi.org/10.1016/j.neuron.2013.06.028

Munn BR, Müller EJ, Aru J, Whyte CJ, Gidon A, Larkum ME, Shine JM (2023) A thalamocortical substrate for integrated information via critical synchronous bursting. Proc Natl Acad Sci USA 120(46):e2308670120. https://doi.org/10.1073/pnas.2308670120

Nawreen N, Baccei ML, Herman JP (2021) Single prolonged stress reduces intrinsic excitability and excitatory synaptic drive onto pyramidal neurons in the infralimbic prefrontal cortex of adult male rats. Front Cell Neurosci 15:705660. https://doi.org/10.3389/fncel.2021.705660

Perez-Cruz C, Müller-Keuker JI, Heilbronner U, Fuchs E, Flügge G (2007) Morphology of pyramidal neurons in the rat prefrontal cortex: lateralized dendritic remodeling by chronic stress. Neural Plast 2007:46276. https://doi.org/10.1155/2007/46276

Powers A, Fani N, Cross D, Ressler KJ, Bradley B (2016) Childhood trauma, PTSD, and psychosis: findings from a highly traumatized, minority sample. Child Abuse Negl 58:111–118. https://doi.org/10.1016/j.chiabu.2016.06.015

Radley JJ, Sisti HM, Hao J, Rocher AB, McCall T, Hof PR, McEwen BS, Morrison JH (2004) Chronic behavioral stress induces apical dendritic reorganization in pyramidal neurons of the medial prefrontal cortex. Neuroscience 125(1):1–6. https://doi.org/10.1016/j.neuroscience.2004.01.006

Ramkumar K, Srikumar BN, Venkatasubramanian D, Siva R, Shankaranarayana Rao BS, Raju TR (2012) Reversal of stress-induced dendritic atrophy in the prefrontal cortex by intracranial self-stimulation. J Neural Transm (Vienna) 119(5):533–543. https://doi.org/10.1007/s00702-011-0740-4

Shansky RM, Morrison JH (2009) Stress-induced dendritic remodeling in the medial prefrontal cortex: effects of circuit, hormones and rest. Brain Res 1293:108–113. https://doi.org/10.1016/j.brainres.2009.03.062

Suzuki M, Larkum ME (2020) General anesthesia decouples cortical pyramidal neurons. Cell 180(4):666–676.e13. https://doi.org/10.1016/j.cell.2020.01.024

Tillmon H, Soteros BM, Shen L, Cong Q, Wollet M, General J, Chin H, Lee JB, Carreno FR, Morilak DA, Kim JH, Sia GM (2024) Complement and microglia activation mediate stress-induced synapse loss in layer 2/3 of the medial prefrontal cortex in male mice. Nat Commun 15(1):9803. https://doi.org/10.1038/s41467-024-54007-5

Tononi G, Boly M, Massimini M, Koch C (2016) Integrated information theory: from consciousness to its physical substrate. Nat Rev Neurosci 17(7):450–461. https://doi.org/10.1038/nrn.2016.44

Urban P, Rezaei V, Bokota G, Denkiewicz M, Basu S, Plewczyński D (2019) Dendritic spines taxonomy: the functional and structural classification • time-dependent probabilistic model of neuronal activation. J Comput Biol 26(4):322–335. https://doi.org/10.1089/cmb.2018.0155

Woo E, Sansing LH, Arnsten AFT, Datta D (2021) Chronic stress weakens connectivity in the prefrontal cortex: architectural and molecular changes. Chronic Stress (Thousand Oaks) 5:24705470211029254. https://doi.org/10.1177/24705470211029254